# IS MY COFFEE CONSCIOUS?

**FOR THE PURVEYORS OF KILLAY'S FINEST COFFEE, ON WHOSE PREMISES MOST OF THIS BOOK WAS WRITTEN.**

First published in Great Britain in 2026 by Cassell, an imprint of
Octopus Publishing Group Ltd
Carmelite House, 50 Victoria Embankment, London EC4Y 0DZ
www.octopusbooks.co.uk

An Hachette UK Company
www.hachette.co.uk

The authorized representative in the EEA is Hachette Ireland,
8 Castlecourt Centre, Dublin 15, D15 XTP3, Ireland (email: info@hbgi.ie)



Distributed in the US by Hachette Book Group
1290 Avenue of the Americas, 4th and 5th Floors, New York, NY 10104

Distributed in Canada by Canadian Manda Group
664 Annette St., Toronto, Ontario, Canada M6S 2C8



ISBN: 978-1-78840-655-0
eISBN: 978-1-78840-656-7

A CIP catalogue record for this book is available from the British Library.

Printed and bound in China.

10 9 8 7 6 5 4 3 2 1

Additional picture credits:
Cleveland Museum of Art/Gift of Ralph King 1917.42; iStock/Anna_leni, appleuzr, Ashva73, Bakai, Barks_japan, bicara tentang brand, bubaone, chocolat, cnythzl, d-l-b, fonikum, Giorgi Gogitidze, GrafikLab, grmarc, Hasanudin Marzuki, ilyast, imdproduction, Invision Frame Studio, JakeOlimb, leremy, Liudmila Baskelovich, Maksym Drozd, MeggiSt, M-Vector, Olena Poliakevych, RaulAlmu, Selahattin Volkan Kuru, Serhii Borodin, SirVectorr, Stockyarder, StudioBarcelona, Sudowoodo, TukTuk Design, ~Userba9fe9ab_931, Vector designer, VectorSilhouettes, viktor kashin, words potrayal; Shutterstock/BrightRainbow, IRIN2000, Unique Design Team

Commissioned by Stephanie Selçuk-Frank
Senior Editor: Alex Stetter
Art Director: Yasia Williams
Designer: Clare Huntley
Illustrator: Joe McDermott
Picture Research Manager: Jen Veall
Senior Production Manager: Peter Hunt

MIX
Paper | Supporting responsible forestry
FSC www.fsc.org FSC® C008047

# IS MY COFFEE CONSCIOUS?

GARETH SOUTHWELL

## PSYCHOLOGICAL ANSWERS TO LIFE'S KEY QUESTIONS

CASSELL ILLUSTRATED

# CONTENTS

# INTRODUCTION

**For newcomers to the subject, it's quite hard to describe what philosophy is – for that, in itself, is a philosophical question. In addition, there's nothing that philosophers like more than to disagree about the definition of their own subject. However, if herded into a room and goaded with tasers, they might reluctantly agree that philosophy is the process of seeking answers to the deepest questions.**

But doesn't science do that? Or religion? Well, yes, to an extent. But the sorts of answers philosophy seeks are subtly different. A physicist might seek to understand how the Big Bang happened, while a theologian might seek to understand how we fit into God's plan. But a philosopher will ask whether it makes rational sense to imagine that the whole universe expanded out of nothing in one massive event. Or they might question God's existence, asking whether it makes sense to say that God gave us free will, when He knows in advance all that will happen.

This isn't to say that philosophers are party-poopers, intent on pricking everyone's balloons (though some are). Rather, it's that, when the scientist and the theologian have done their sums, scrutinized their scriptures and are sitting with their feet up, drinking a nice cup of tea, the philosopher is the annoying, insistent child tugging at their shirtsleeves, still asking, 'Yes, but *why*?'

(As I write this, I can sense the itchy fingers of fellow philosophers – and probably some scientists and theologians, too – hovering over their keyboards, waiting to correct me. But I'd be disappointed if that weren't the case.)

A related question you might have is: 'What's the point of philosophy?' Well, I must admit, it does mostly consist of questions and not very many answers. It doesn't necessarily give us comfort on dark, lonely nights – in fact, at times, it does the opposite. It's usually of no use when your toaster breaks, or if you need a plan to get yourself out of debt. In fact, it can often seem to be of very little practical use at all. Fingers crossed you still have the receipt for this book...

That said, practicality isn't everything. And I should say, in philosophy's defence, that some of its theories do have everyday applications.

- **Ethics, or moral philosophy**, is mainly concerned with how you should live your life, what goodness is and how you can be happy.

- **Political philosophy** looks at what form society should take and how we can ensure the rights, justice and freedom of its members.

- **Aesthetics, or the philosophy of art**, looks at how we value creative things and helps us understand how they give meaning to our lives.

- **Philosophy of mind** asks what we essentially are and whether we need to worry that we'll be replaced by AI.

- **Epistemology, or the theory of knowledge**, looks at how we know things, and whether and how we can be justified in our beliefs.

So, it's not true to say that the subjects covered in the following pages have no application. What a terrible advert for this book that would be.

However, if I'm honest (and Immanuel Kant, the German Enlightenment philosopher, says I must be), philosophy is really for thoughtful types. For those who've ever wondered about unusual things and weren't sure what to do with their ponderings. It's for the curious, the reflective, the weird. It's for those who've ever wondered if they really woke up this morning and aren't still dreaming. Or whether they can ever really know what true love is. Or if shopping can make them happy. Or what comes after death, and possibly before birth. It's for anyone who's ever wondered – to the annoyance of their friends and family – *why*.

It's for you.

# ME MYSELF AND I

## THE MIND

# 1.

# WAS I NAPOLEON IN A PAST LIFE?

**In an idle moment, who hasn't wondered if death is the end? Might life continue on in some disembodied form? But if that's a possibility, then might it not also be the case that birth was not the beginning? In other words, might it be possible that you have lived before?**

When you start thinking along these lines, it's easy for your imagination to run away with you. Perhaps you were someone famous. Napoleon, or Marie Antoinette. It's true that you are very fond of croissants. Given that most religious traditions hold that the soul has no gender, perhaps you were someone of the opposite sex. This makes Cleopatra as likely, or unlikely, as Shakespeare. Of course, you shouldn't ignore the possible downside – that you were someone you might not *want* to remember being (and history would prefer to forget...).

Before you get carried away with fantasies of this kind, is there any good reason to hold such a belief? Reincarnation is an old idea and is a central tenet of many faiths, including Buddhism, Hinduism and Sikhism. Although this means it's often associated with Eastern thought, reincarnation can also be found in Ancient Greece at the very beginnings of Western philosophy, with thinkers such as Plato (427–347 BC) and Pythagoras (*c*.570–*c*.497 BC).

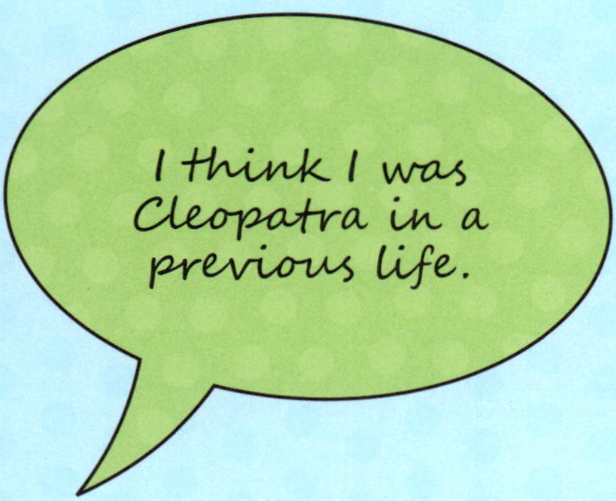

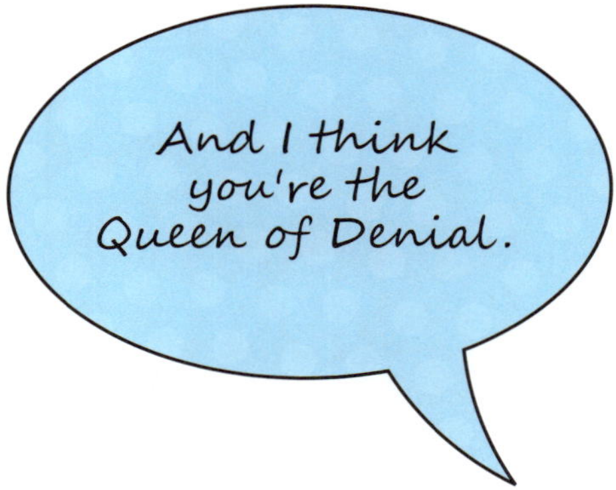


In his philosophical dialogue the *Phaedo*, Plato sets down several arguments for the existence of the soul and its continuation post-mortem. In the dialogue, which is a sort of play, a fictionalized version of Plato's mentor Socrates (469–399 BC) sits awaiting execution by poison. He passes the time debating with his friends and followers, and tries to reassure them that death is not the end. For instance, Socrates argues that the reason we know certain unquestionable truths in this life is because we were born with innate knowledge. Therefore, we must have existed in a disembodied state before birth and had first-hand acquaintance with such eternal truths. As such, all learning is really only recollection of this pre-birth knowledge.

In another argument, Socrates concludes that the soul is distinct and opposite in nature to the body, and since we know from experience that the body is mortal and subject to decay, then we must conclude that the soul, being divine, is immortal and both predates and survives the body.

However, neither of these arguments is particularly strong. We shall return to the question of innate knowledge later in the book (see page 78), but we don't have to conclude that fundamental truths must have been learned by a pre-existing soul in a disembodied state; they might simply be assumptions that we have acquired through experience, or represent the necessary order our brains impose in order to make sense of reality. When it comes to the distinct nature of the body and soul, this first assumes that there is indeed something distinct from the body called the soul. We could equally argue that what Plato calls the soul is simply

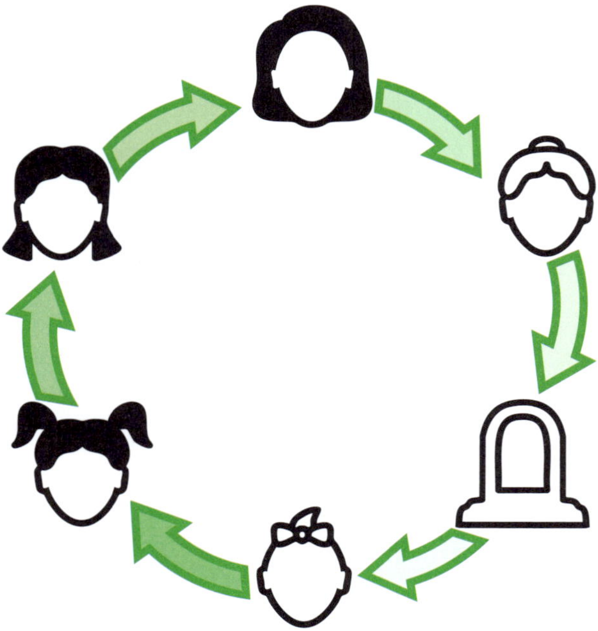

CYCLE OF REBIRTH

the result of natural physiological processes in the brain (although, as we'll see later on, this view comes with its own conundrums...).

In another dialogue, the *Timaeus*, Plato states that each of the stars is associated with a soul. Since the number of stars was thought to be finite, this seems to suggest that the number of souls were too, and highlights another problem for believers in reincarnation. The population of the world in the time of Plato was something like one hundred and fifty million people; it is now just over eight billion. Where have all those extra souls come from? Are there queues of pre-made souls in some spiritual dimension, waiting to be born? And if so, what decides which ones get their first go, and which ones are given another turn?

Of course, even if Plato is wrong in the details and his arguments are flawed, he may not be wrong in principle. For instance, it could be that new souls can come into existence from some pre-existing store of 'soul stuff', and when people die their souls are simply reabsorbed into it, as puddles on the shore are reclaimed by the sea. It might also be that, rather than being individual and distinct, souls share the same underlying spiritual essence.

But this is all conjecture and, ultimately, subsequent philosophers have found Plato's arguments unconvincing. This is not to say that it is irrational to hold a belief in reincarnation. In fact, there are many intriguing cases of so-called past-life memories and near-death experiences that merit investigation. A well-known collection of the latter is Dr Raymond A Moody's 1975 book, *Life After Life*. However, since such examples seem to require trust in the authenticity of others' reports, and can potentially be explained in other ways, then the last step in belief in reincarnation would seem to involve faith and personal conviction rather than conclusive proof.

> As a man discards his threadbare robes and puts on new, so the Spirit throws off Its worn-out bodies and takes fresh ones.
>
> Bhagavad Gita, chapter 2, verse 22.
> Translation by Shri Purohit Swami

# AM I JUST MY BRAIN... OR SOMETHING MORE?

**What are you? A human being, certainly. In addition to that, a person of a certain age and sex, nationality and ethnicity, weight and height. You also, no doubt, possess certain personal preferences: perhaps a penchant for salsa dancing, a liking for pistachio ice cream or a fondness for reading whodunnits. But are any of these the real you? Tastes change, weight fluctuates and you grow then shrink with age. On the other hand, 'a mostly hairless bipedal mammal' doesn't really nail anything essential about you as a person. So, what's the answer? Is there one?**

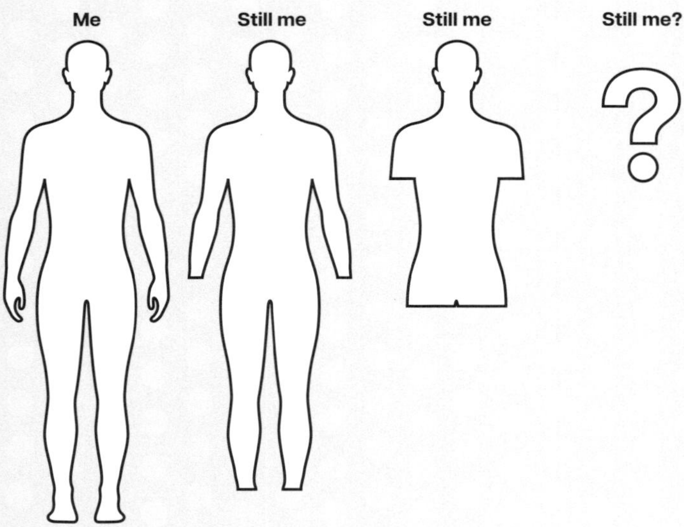


WITHOUT MY BODY, WOULD I STILL BE ME?

Sometime during the first quarter of the 17th century, similar questions to these occurred to a young French philosopher and mathematician by the name of René Descartes (1596–1650). Disturbed by various sceptical arguments current at the time, he began to ask himself if there were anything of which he could be certain. Surely, he reasoned, even if my knowledge of the world were faulty, I could at least be certain that I exist – couldn't I? For to doubt that would be madness. But what then *am* I?

In answering this question, Descartes concluded that if there is anything about 'me' that I can conceive of myself as existing without, then it is not essential to what I really am. So, for instance, I can conceive of myself without fingers or toes, arms, legs or various other body parts, for their loss would not render me any less 'me'. But what about the body itself? If I really push the boat out, sceptically speaking, could I conceive of myself as existing without a body at all?

Descartes concluded that he could. One reason for this is that all that I know about my physical self may be called into doubt. For really, what do I know about my body anyway? Only what I learn about it through my senses – by touch, sight, smell and so on. But my senses don't really have a great track record in terms of trustworthiness. Descartes notes times in the past where his senses misled him. We can all think of examples here – a tree in the distance that looks like a man; a plastic bag hoisted by the wind that contrives to look like a bird – and modern psychology has since had a giddy time cataloguing the myriad ways in which judgements based solely upon sensory information are often just plain wrong.

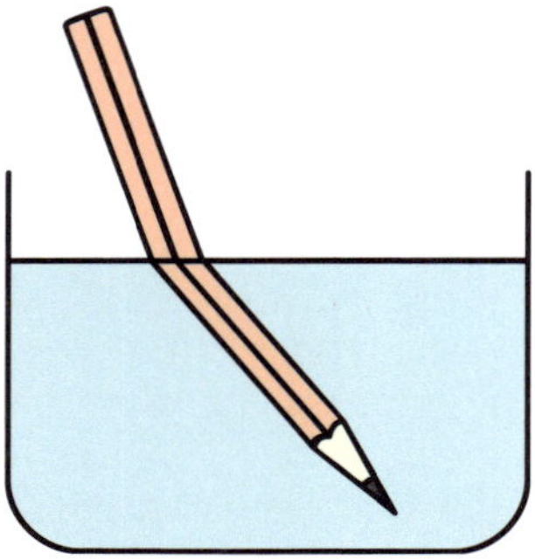

OPTICAL ILLUSION

We'll come back to this later in the book when we look at the problem of scepticism (see page 85), but here I want to focus on Descartes's answer as it relates to the question of what I am. I can, he said, doubt everything about my existence, except for the fact that I think. This is the basis of what has become known as his *Cogito* argument, from the Latin phrase *Cogito ergo sum*, 'I think, therefore I am'.

Let us suppose that what I think of as the real world is an illusion, that I am living in a dream or some manufactured virtual environment. Bizarre as the possibility seems, it might even be that only I exist (a view that is known as solipsism). If something like this were true, then even what I think of as my body might just be a figment of my imagination. But even if this were the case, I couldn't be fooled into thinking that I don't possess a mind. For to be fooled, to doubt, to make false judgements or draw faulty conclusions, I must be thinking, as all of these actions require thought. And so, even in the act of being mistaken about absolutely everything else, I cannot be wrong about the fact that 'I' – as a conscious, thinking being – in some sense exist.

To return to the question of what I really am, Descartes would therefore answer that I am a thinking thing. I'm not a salsa-dancing thing, a pistachio-ice-cream-eating thing or a whodunnit-reading thing – or at least, not essentially. I am first and foremost a thing that thinks; the rest comes later. But what then is this thing? What is the vehicle for this activity of thinking? And here, as a devout Catholic, Descartes's own religious convictions come in, as he believed that this thinking essence was none other than the Christian soul. I have a body, but I am not one; I am a mind or soul. For this reason, Descartes's view is known as Cartesian dualism (from *Cartesius*, the Latin form of Descartes), as it proposes that the universe is dualistic, being made up of two substances: mind and matter.

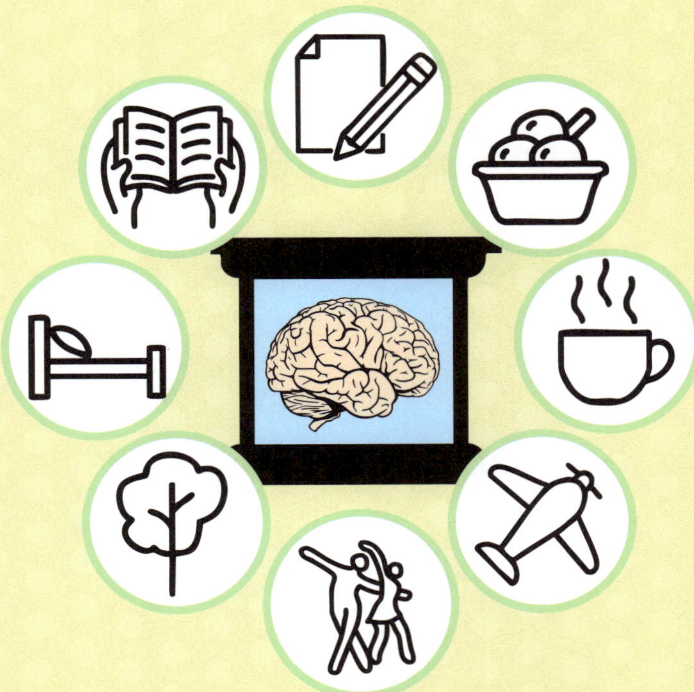

HOW DO I KNOW IF MY WORLD IS REAL?

However, Descartes's *Cogito* does not have to be interpreted religiously and his key insight here is that consciousness itself cannot be an illusion. This, in fact, continues to cause problems for modern philosophy and science, which mostly rejects Descartes's dualistic and religious view in favour of physicalistic monism, which proposes that only one type of stuff exists, and that is physical. But even for this view, it remains a conundrum how the atoms that make up the brain can produce mental experience. How does insentient physical stuff become sentient? How does quantity become quality? Whatever the answer, and whatever the true nature of the universe might turn out to be, it remains a fundamental truth that conscious, thinking beings exist in it.

Well, one of them, at least.

# ARE MY MEMORIES REALLY MINE?

**You wake up in a hospital bed with no idea of how you got there. There's an unfamiliar woman sitting by your side, who says she's your wife and tells you that you've suffered a traumatic head injury from a road accident, which has affected your memory. But don't worry, she says, the doctors are confident that it will all come back to you. And, gradually, it does: your life as a science-fiction author; your semi-detached house in the suburbs; your two beautiful children, Doris and Dave. And so, you slowly reacclimatize to your old habits. Life is good again.**

Or is it? One day, another strange woman shows up at your door, claiming that she is your wife, that there was no accident and that you are, in fact, a computer scientist working on a top-secret project for the government. You've been kidnapped by enemy agents who have neurochemically implanted you with false memories, and in the sci-fi story you're currently writing, you're unwittingly spilling all your top-secret research.

Does any of this sound plausible? Not the memory-implant part or the enemy agents, but the idea that memory is essential to who you are – or as philosophers term it, your personal identity – and all that's needed for you to think yourself someone completely different is for your memories to change.

The above scenario is similar to the plot of the 1990 film *Total Recall*, which was based on the 1966 novelette *We Can Remember It for You Wholesale*, by the science-fiction writer Philip K Dick. In the film, office-worker Douglas Quaid one day decides to take a 'mental holiday' by signing up for a memory implant procedure that will convince him that he has been on an exciting mission to Mars as a secret agent. During the procedure something goes wrong and Quaid becomes convinced that he's a real secret agent. We're then

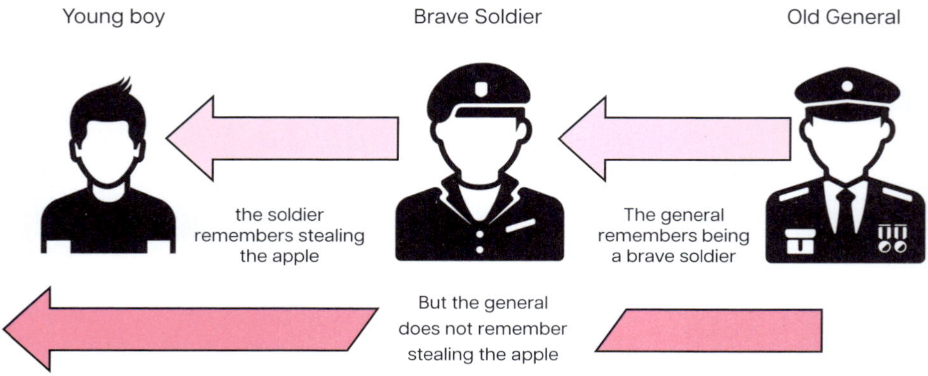


REID'S OBJECTION TO LOCKE

left wondering if Quaid is a secret agent who is remembering his true identity, or someone who has mistaken a false, implanted memory for the truth.

The idea that memory is central to personal identity was first proposed by the English philosopher John Locke (1632–1704) in *An Essay Concerning Human Understanding* (1689). Locke argued that if you can recall some experience from the past in the same way that you are now experiencing the present, then you are the same person in both cases. This all seems obvious: you remember playing golf yesterday (the swing of the 9-iron in your hand and the breeze on your face), therefore you played golf yesterday. However, there are several problems with this theory.

The first objection was proposed by Scottish philosopher Thomas Reid (1710–96), who pointed out that there are often cases where we don't remember what we did. For example, do you remember what you did three Mondays ago? Does that mean that we're not the same person as the one who had the experience?

Reid illustrates this problem with the example of a boy who was caught stealing an apple. The boy went on to be a brave soldier and later became a general. As the general ages, he forgets that he stole the apple as a boy but remembers being the brave soldier. When he was a brave soldier, he remembered stealing the apple as a boy. What results when we put these statements together is a contradiction or a paradox: according to Locke's criterion of memory, the general was the soldier and the soldier was the boy, but the general was not the boy. In case you're wondering, paradoxes in philosophy are generally considered a bad thing and a sign that something's gone horribly wrong with a theory.

Another challenge to Locke's idea came from English philosopher Bishop Joseph Butler (1692–1752), who argued that the theory is circular. In other words, Locke might have a memory of playing golf yesterday and so concludes that he was the person who played golf yesterday, but in doing so he assumes that this is a genuine memory. As my kidnapped scientist scenario and *Total Recall* both highlight, the fact that we can remember doing something is not itself proof that the memory is true, or even ours, and so it cannot act as proof that we are the same person as the one who supposedly had that experience.

You might think all of this is overblown. We know that memory can be untrustworthy and it's quite common for us to forget details even about what we did last week. The fact is, what makes us the same person is having the same body; the boy is the same person as the general because the former grew up to be latter, whether or not the general later remembers anything about his boyhood.

However, as we'll see, there are problems with this view too.

# SHOULD I BE HAPPY IF MY CLONE SURVIVES ME?

**It's 2069. Humanity has begun to colonize the solar system and there are now human settlements on Mars, Earth's Moon and one of the moons of Jupiter. What has facilitated this process is an amazing new technology called the Xuber. This allows your body to be scanned and the resulting information sent to a distant planet, where it's installed in a new physical brain and body. This clone is an exact replica of your body in its current state, even down to the bruise you got when you bumped your arm on the banister yesterday. To avoid legal and social complications, once the interplanetary transfer is made, the old you is immediately destroyed.**

You're about to go on holiday to Mars, where you'll join your family members who have already made the transfer. They've used the technology before, but it's your first time and you're a bit nervous. As you sit in the Xuber, something goes wrong and, rather than passing into unconsciousness and waking up on Mars, you find yourself in your old body, still on Earth. An artificial voice informs you that due to a technical error, your body has not been destroyed, although the faulty procedure has delivered fatal radioactive damage to your cells. In a day or so, you'll be dead. But don't worry, everything is fine! You chat with your family by interplanetary video call and the new you waves back at the old you – you even see that bruise on your arm. Your wife and two children are standing beside the new you. They seem happy. Shouldn't you be too?

This scenario is slightly adapted from one explored in various ways by the English philosopher Derek Parfit (1942–2017) in his 1984 book *Reasons and Persons*, as he imagines the potential effect of such a teleportation device on our notion of personal identity. We've seen how the idea that my memories make me the same person from moment to moment can run into trouble under certain circumstances – such as implanted false memories or memory gaps that

appear through ageing. You might think that a simpler solution is to base our personal identity on the fact that we are the same *physical body* from moment to moment. But there are problems here, too.

First, there's a scientific sense in which you are *not* the same body from moment to moment. This is illustrated by the problem of Theseus's ship. In Ancient Greece, the famous ship in which the hero Theseus returned from Minoan Crete after defeating the Minotaur was said to be preserved in the port of Athens. Over the years, the planks, ropes and timbers rotted and were replaced, one by one, until eventually, there was not an original piece in the whole ship. At which point, can we really say that it's the 'same' ship as it originally was?

This scenario can also be applied to the human body. Your cells are dying off and being replaced every second, even your atoms are being replaced by new ones, and so over a period of years, arguably you're a completely different person, physically speaking. Does that mean that you are no longer you?

Maybe this isn't an issue – perhaps we can say that you're the same body, because there are physical processes over time linking the bits you now are and the bits you once were. However, we can imagine other problematic scenarios.

OVER TIME, EACH CELL IS RENEWED

Let's say that some Dr Frankenstein type transplants half your brain into someone else's body, and half of that other person's brain into yours. Which body is now you? Or, given that – arguably – our uniqueness lies in the memories, experience and information physically encoded in our brain, could these not be scanned and transferred to a completely fresh body (as with our Martian teleportation example)? Or even stored in a computer? And in that case, someone might make copies of your personality, and there could be more than one you.

There are many conclusions to draw from Parfit's teleportation example, but concerning the idea that the person is just the body (or more specifically, the brain that stores the personality), it demonstrates that under certain, admittedly far-fetched, scenarios, there is no such thing as a unique you. Even if we could patch up the problems inherent in Locke's memory-based theory of personal identity (and

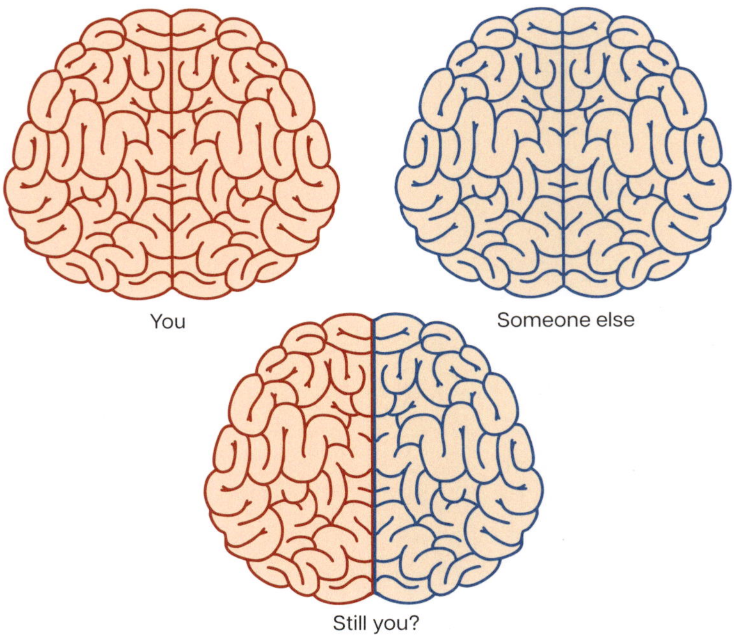


BRAIN SWAP EXPERIMENT

Parfit tries his best), what we have is still subject to the sort of brain transfer/transplant/clone/digital copying scenarios beloved of science-fiction writers. Does that mean – psychologically and physically – that there is no concept of an essential self that could surmount all of these problems? Parfit thinks not. But maybe this isn't all bad news, for at least this means that, whatever happens, you will survive in some form – perhaps more than one. In which case, as you lay dying on Earth, you should perhaps be happy that your clone will outlive you – because they are, to all intents and purposes, as close a thing to you as will exist thereafter.

That said, this is probably a depressing conclusion for most of us, as we might feel instinctively that it isn't the real me that lives on. However, Parfit asks us to take consolation in the fact that, when we die, there is some part of us that lives on in the memories and lives of others, the things we once owned, the atoms that once made up our body and the projects we set in motion. Together, these still make up a part of your self in a bigger sense. Granted, these things are not conscious in quite the same way as you once were and you may still feel that this 'liberation from the self', as Parfit calls it, is not really you in any way that matters. But maybe that's better than nothing?

Now, where did I put those books on reincarnation…

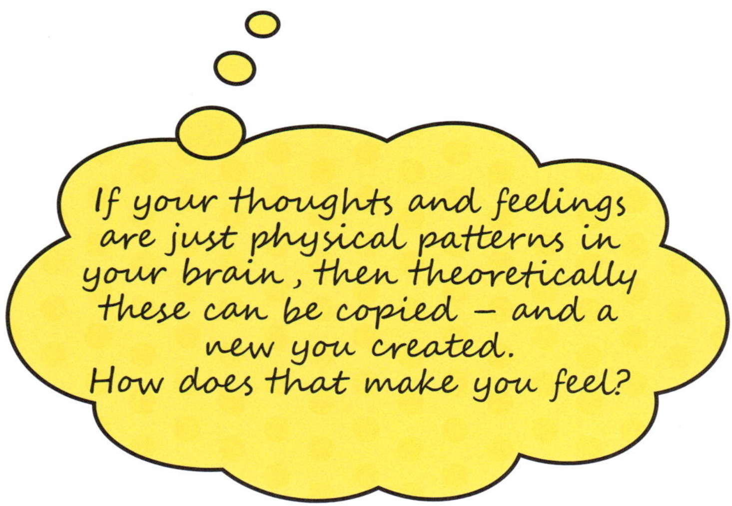

# CAN A MACHINE TRULY THINK?

**Computers have come a long way in the last 50 years or so, especially in recent times, during which artificial intelligence (AI) has made giant leaps forward. On the back of this, some computer scientists, philosophers and AI enthusiasts have suggested that it is only a matter of time before there will be computers that can do everything a human can, intellectually speaking – what's called artificial general intelligence (AGI) – and even beyond. Some people have even gone so far as to claim that machines will become sentient, having their own thoughts, perceptions and sense of self, and deserving 'machine rights' to parallel those we accord to humans.**

However, there are good reasons to think that some of this enthusiasm is at least premature, while some of it is completely misguided.

This question was explored by mathematician Alan Turing (1912–54) in his 1950 paper 'Computing Machinery and Intelligence'. In it, Turing argued that all that is needed for us to conclude that a computer is thinking is that it can fool an objective observer into believing that it is human. This is the famous Turing Test, or as Turing himself called it, the imitation game.

The test involves a situation where a human judge asks questions of a machine and another human, respectively, and must then decide, based on written communication alone, which is human. If the machine can imitate human responses so well that the interrogator is fooled, then the machine has passed the test: it is thinking, to all intents and purposes. Arguably, programmes like ChatGPT can already carry on such conversations convincingly, even using artificial voice or video generation. Does that mean the Turing test has been passed?

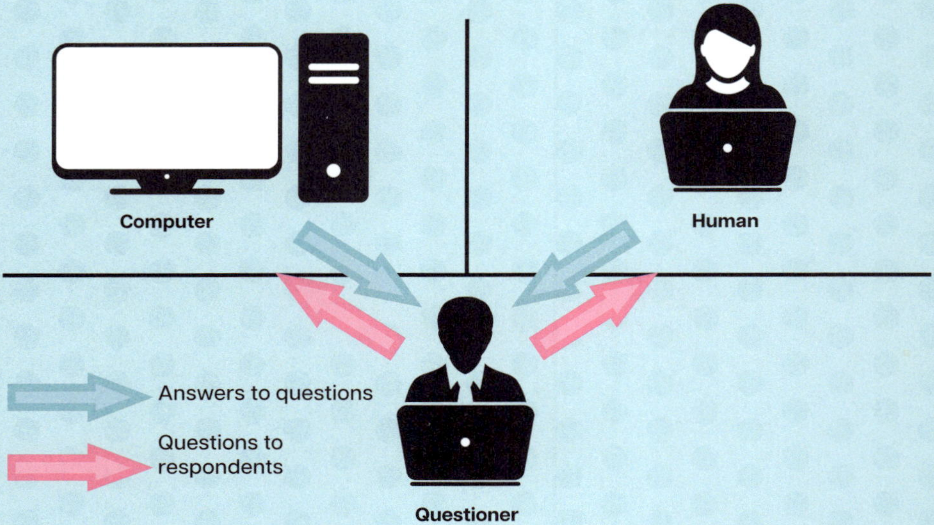


THE TURING TEST

There are various problems with Turing's test, not least of which is whether the human propensity to be fooled should be proof of anything. It has also been questioned whether the functions that machines are capable of can ever be considered thinking in the fullest sense. Physicist and mathematician Sir Roger Penrose (1931–) has argued that, while computers operate via a determinate set of rules or algorithms, the human mind does not. This means, potentially, that there are insights a human mind could achieve that a computer could not.

An example of this is the so-called halting problem, which asks whether we can design a programme that would be able to decide whether another programme would halt (finish its programme), or continue for ever. The specifics of this are quite complicated, but the gist, according to Turing himself, is that this is impossible for all cases. Penrose takes this conclusion and uses it to argue that human intelligence is not built solely on rules and there are certain types of logical and mathematical problems that a human could answer but a machine could not. Therefore a machine can never truly think in the fullest sense.

Whether this is true or not, there are other reasons to think that computers and humans are fundamentally different in terms of their capabilities. Philosopher John Searle (1932–2025) has argued that a computer can never understand something in the way that a human mind can because it's a rule-driven system and it's possible to follow rules without understanding them.

Searle illustrates this with what has become known as the Chinese room thought experiment. Imagine there's a person sitting in a small room with two hatches. Tiles with Chinese symbols on them are delivered through one hatch. As they arrive, the person looks up the symbols in a book. The book does not tell them what the symbols mean, it only instructs them which tiles to select from a stack next to them in response. The person then deposits the selected tiles out through the other hatch.

Could we say, in this scenario, that the person understands Chinese? Searle argues that we could not, because all they would be doing is following the rules in the book (the syntax

'Penrose argues that while a human could see that a set of commands will stop or run for ever, a machine cannot in all cases.'

THE HALTING PROBLEM

of the language), while remaining ignorant of the symbols' meaning (the semantics). The same conclusion applies to computers, which are similarly just syntactical machines unconsciously following rules. AI, Searle is saying, can never understand in the fullest sense, because there is no process whereby anything can mean anything to it.

So, how do humans differ? For Searle, it comes down to consciousness. Words have meaning for us because they relate to how we think and feel, reflecting our goals and intentions. If I translate a sentence from Chinese and I truly understand it, its meaning might make me blush or get angry or sad. But this is more than simply following a rule, such as: 'When faced with this sentence, engage blushing circuits.' It requires a feeling of embarrassment, which is something a machine can never experience.

# IS MY COFFEE CONSCIOUS?

**As you gaze into your latte, does your latte also gaze back into you? This might sound like the sort of silly and outrageous claim designed to make you stop and click on a website post, pick up a magazine or maybe even buy a book, but as with most crazy-sounding things in philosophy, there's a serious question behind it.**

The problem of consciousness is one of the toughest and most controversial in philosophy. How does physical matter give rise to thoughts and feelings? How do all those neurons, synapses and axons result in the smell of coffee, the taste of strawberries or the redness of a rose? These subjective experiences – what philosophers call *qualia* – seem to have no place in the hard-headed objective picture painted by physicists and neuroscientists. So how then do we account for them? With great difficulty, it seems, for which reason it has been christened by Australian philosopher David Chalmers (1966–) 'the hard problem', to be distinguished from the comparatively 'easy' scientific task of working out how the brain itself works.

Historically, there have been two main responses to the problem. We've already touched upon one, proposed by Descartes (see page 19), who argued that the subjective world of impressions, thoughts and feelings, and the objective world of weights, shape and size, correspond to two different substances. For this reason, his view is known as substance dualism. But the problem with this, as already noted, is how such different substances interact. If they are so different, then how can a thought 'nudge' a neuron, or a neuron provoke a thought?

For these and other reasons, dualism fell out of fashion, to be replaced by the simple view that now pervades: matter is all that exists (materialism), and the mind is simply the brain (what's called identity theory). But there are problems with this view, too. First, in what way are brain states identical to mental states? To say, 'I'm going to the shops on

Tuesday' is not in any clear way the same as saying 'these particular neurons are firing'. The language we use to talk about the mind and ourselves is of a different sort to that used by neuroscientists to talk about the brain, and if the terms don't correspond, then how can there be identity between them?

An answer to this is to say that our everyday way of speaking about ourselves is simply wrong. Just as superstitious folk-beliefs have, over time, been replaced by scientific understanding, so our notions about ourselves, our mistaken understanding of our 'thoughts', 'feelings' or 'desires', will one day be eliminated in favour of proper neuroscientific concepts and language, in an approach known as eliminative materialism. But aside from the fact that the way we speak about ourselves isn't meant to be a theory that can be proven right or wrong, this solution does nothing to address the problem of consciousness itself. Are we simply mistaken about that, too?

Astonishingly, there are philosophers who believe that consciousness itself is an illusion (a position aptly named illusionism). This second approach would indeed solve the hard problem, but at the expense of dispensing with something that most of us consider the most fundamental truth that we have: that we are conscious beings. We might also point out that to have a mistaken experience is one thing, but to be mistaken as to the existence of experience itself involves the very thing it wants to disprove. I can mistake a distant tree for a person, but I can't be wrong about *having* an experience (whether illusory or not). To be mistaken, I must be conscious. Even a sensory mistake is proof that I am a conscious being and that I exist, which was the conclusion that Descartes himself arrived at.

But maybe we've been going about the hard problem all wrong. In *Galileo's Error* (2019), philosopher Philip Goff (1978–) argues that there is a neglected third way, which is called panpsychism. This is not a new theory, but it is regaining respectability in the light of the ongoing difficulties that materialists face in explaining consciousness. Goff argues that, in focusing solely on measurable, objective qualities, science has neglected to factor in the possibility that consciousness is something that was simply built into the universe from the start. So, rather than attempting to solve the problem of how the physical brain produces subjective mental states, we simply assume that consciousness is what Goff calls 'the intrinsic nature of matter' – namely, what it is like to be a physical thing.

However, this view comes with its own problems. Firstly, we can't ring-fence those things that most people concede are conscious or sentient – people, animals, insects, perhaps plants – but must also allot consciousness to amoebae, viruses and DNA. So, without meaning to put you off your cappuccino, can you imagine the mental states of the tiny bacteria-like creatures that probably live on in your milk foam and your dusting of chocolate powder? And not only organic things, but inorganic things too. Does it feel like something to be a stone, a stream or an electron?

Of course, panpsychists don't think there is much subjective life going on inside a rock. The point is, rather, that if there are building blocks of matter, then why not of consciousness as well? But this too presents a difficulty (called the combination problem), because how do all these separately conscious atoms combine into a unified mind? How do all the conscious bits form a conscious whole? Although, given that the human mind is modular (built up of distinct lobes and functional areas assembled into a single system over the course of evolution), then perhaps this isn't the problem that panpsychism's critics think it is.

Whatever the case, panpsychism would certainly solve a lot of philosophical problems and would help us reconnect with a living world of which we are an integral part, rather than the inanimate and mechanical universe bequeathed to us by Galileo and modern science. Consciousness is not some mysterious thing floating around, disconnected from the world, but something inside it. Consciousness is, in fact, what it is like to *be* the world.

WHAT'S IT LIKE TO BE AN ANIMAL, AN INSECT, A ROCK?

TV+
N
luhu
CINEMA CLASSICS
WILD WESTERNS
SCI-FI
MAX
COMEDY
GLOBAL NEWS
HORROR

# DO I CHOOSE WHAT I WATCH ON TV?

**You settle down after a long day's work to watch something on TV. But there are so many options, now – so many channels, on-demand streaming services, films and series to rent, buy or watch for free – by the time you finally decide, it feels like a hard-won battle. Unless, of course, you were always destined to make that choice.**

The idea that our every action is pre-ordained, and that choice is an illusion, is known as determinism, and it can take various forms. The broadest is causal determinism, or the idea that, since every physical effect has a prior cause, then everything that happens is fixed in advance. And because, as most scientists argue, the mind is ultimately reducible to the brain, then your every attitude, opinion or desire is the result of some previous physical state of your neurons. And so, your TV choice can ultimately be traced back to the Big Bang – perhaps.

There are also more specific forms of determinism. A biologist might argue for genetic determinism, where all our actions are ultimately the result of our genes. You didn't choose to watch a romantic comedy but were genetically predisposed to favour them by the evolutionary drive to procreate. German political philosopher Karl Marx (1818–83) argued for economic determinism: you choose between a soap opera or a high-brow drama based on your social class, wealth and education, which are in turn dictated by the economic forces that shape society. Austrian father of psychoanalysis Sigmund Freud (1856–1939) argued for psychological determinism, where our choices are shaped by unconscious drives: you choose a film that glamorizes crime because of your unconscious resentment of authority, which in turn springs from your problematic relationship with your father.

However, in all of these more specific forms of determinism, there is still the question of whether we can overcome these influences. If not, they can all be arguably reduced to causal determinism – the physical interaction of atoms, or some other fixed cause – which means that, no matter how hard you try to overcome your social environment, your genetics or your psychological baggage, your success or failure is ultimately out of your hands. Which makes it all feel a bit pointless. If everything is pre-ordained, what's the point in even trying? Even your apathy is a foregone conclusion.

Which is a problem for determinists. If you're not ultimately responsible for your actions, then how can you be censured or congratulated for your behaviour? How can you be judged or punished for your criminality? This absurdity is perhaps most strikingly illustrated by religious determinism: if God is all-knowing, then He knew what you would do even before you were born; which makes life a tragedy or a farce (depending on your viewpoint), created solely for His amusement.

For the above reasons, your average person – and society in general – tends not to believe in strict determinism, but rather that we have at least some free will. This view is known as libertarianism (not to be confused with the political theory of the same name) and proposes that there is some mechanism whereby we may potentially overcome external influences and thereby be responsible for at least some of our actions. The problem with this, of course, is that it seems to require us to reject the so-called 'law of causality', which states that every effect has a fixed cause. Indeed, as we've seen, philosophers like Descartes argued that our free will springs from our possession of an immaterial soul, which is free from physical interference. But aside from the problem of proving the existence of such an entity, there is also the question of how an immaterial substance can influence a material one: how does a non-physical mind influence physical matter?

An alternative, non-religious account of free will might be based on quantum physics. In a famous thought experiment, Austrian physicist Erwin Schrödinger (1887–1961) asked us to imagine a cat sealed in a box with a mechanism that will release a poisonous gas only if a particle is emitted by a radioactive substance. But, as quantum physicists have shown, exactly when that particle will be released is only probabilistically predictable by pure mathematics, not precisely predictable. This means that the only way we can conclusively know whether the cat is alive or dead is to open the box and check. This seemed crazy to Schrödinger, for it suggests that there are no natural laws that fix the outcome, but simply the act of looking itself.

Could this then provide a foundation for free will? Well, no, because the fact that nothing ultimately determines whether the cat is alive or dead (a state of indeterminism) is closer to randomness than freedom, and you would have as little say in a random choice as you would in a pre-determined one.

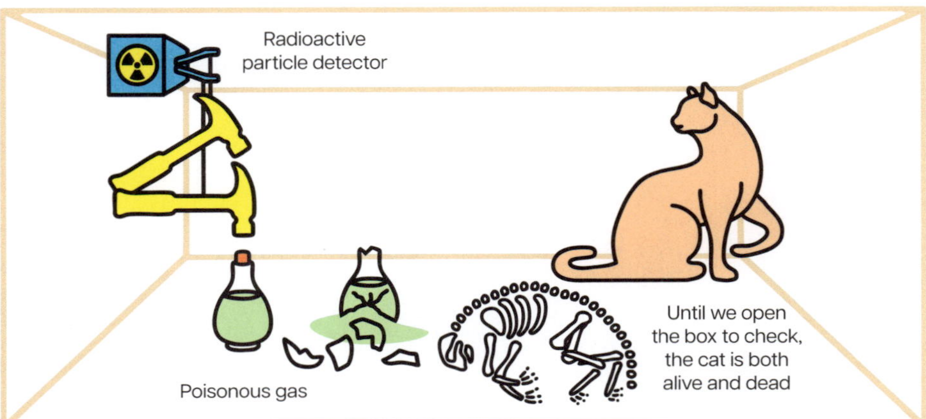


SCHRÖDINGER'S CAT

However, although there are problems with libertarianism, and determinism makes life meaningless, there might be a third way. This course is known as compatibilism, which seems to be the most popular among contemporary thinkers. It was also favoured by Scottish philosopher David Hume (1711–76), who argued that free will is compatible with physical determinism. This is because we can distinguish between those actions that we are traditionally considered to have control over and those that we do not.

So, if I decide to move my arm, to speak or to sign a contract, then those are acts of free will. But I cannot be said to be in control if my leg jerks when the doctor tests my reflexes by tapping my knee with his little hammer, or if I cry out in pain when someone unexpectedly treads on my foot. We might even argue that I am not in control if I am forced to sign a contract under duress. This is not to say that all these actions, whether free or coerced, are not previously determined by the physical laws of the universe, but they are 'free' acts if they are *determined in the right way.*

In fact, to have moral responsibility, our 'free' actions *need* to be determined – for how can we be praised or blamed for actions that have no definite cause? That said, some might argue that compatibilism is merely a relabelling of certain pre-determined acts and doesn't actually give us free will in the deepest, truest sense. Ultimately, we are still biological robots.

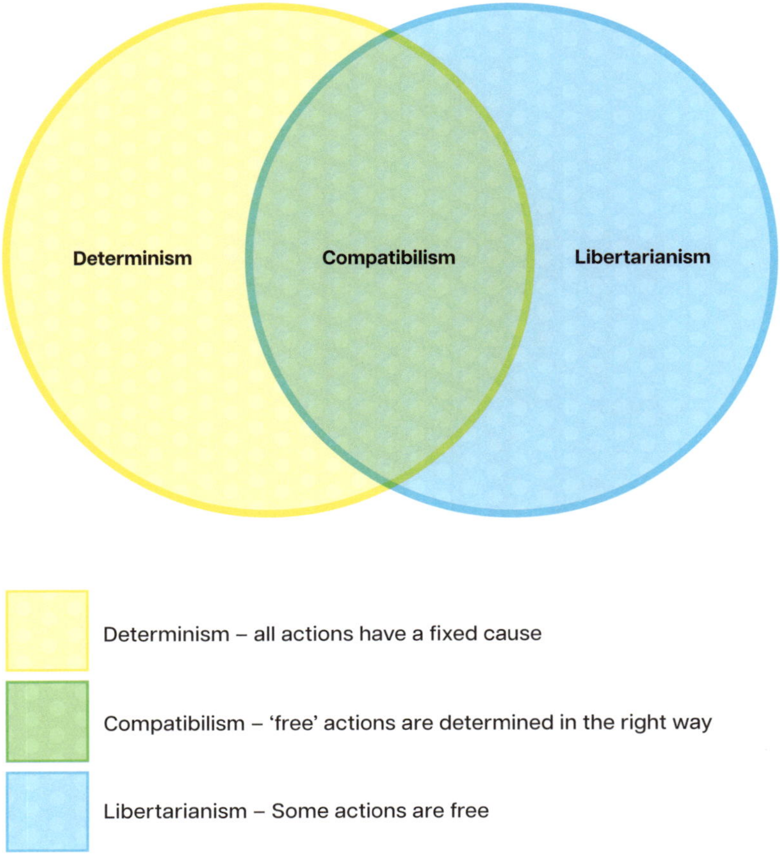


DETERMINISM V FREE WILL

# TRUST ISSUES

## EPISTEMOLOGY AND METAPHYSICS

# 2.

# SHOULD I TRUST WHAT MY TEACHERS TAUGHT ME?

**Our lives are built on trust. When we are very young, we must rely on our parents, not just for survival and emotional support, but for knowledge of how the world works. 'Don't touch that, it burns.' 'Don't eat that, it's poisonous.' 'Don't pull that face, or you'll stay like that.' As we grow older, this parental guidance is partly replaced by education, and the worldview that we graduate to as adults has largely been shaped by our teachers and the society they serve. But can all of this be trusted? Are the facts, values and opinions that society instils in us a reliable foundation for our own knowledge?**

Descartes asked this question of his own beliefs and decided that the answer was no. Admittedly, many things we're told as children turn out to be true, but mixed in among them are superstitions, prejudices, white lies and simple errors. Compare the education of a medieval school pupil with that of a Victorian child, or someone from the early 21st century. Yes, some facts will remain the same, but over time many fundamental things will have changed – to do with the universe, the world, psychology, biology and physics, but also attitudes to social freedom, morality, love and sex.

For his own part, Descartes was primarily focused on scientific knowledge and decided that if we are to arrive at a truer picture of reality, then we need a way to question the beliefs that are handed down to us in order to arrive at things which are beyond doubt, and can therefore act as true foundations for all our knowledge. This process of questioning has therefore become known as his method of doubt.

c Dd Ee Ff Gg Hh Ii Jj
2 + 2 = 4

Descartes's prime candidate for such a fundamental truth is that even if everything else is an illusion, then the fact that I am thinking proves that it is at least true that I exist (the so-called *Cogito* argument, see page 20). This was his foundation stone. But it's not much of one, is it? It doesn't tell him much about the world outside himself. To do that, he reasoned that he must find other ideas that are similar in kind to the *Cogito*, that appear – as he put it – as clear and distinct as the fact that he, a thinking thing, exists.

He had in mind here the sort of basic truths of maths, science and logic: 2 + 2 = 4, every effect has a fixed cause, something cannot come from nothing, all bachelors are unmarried, and so on. It's these sorts of truths that, taken together, can act as a foundation for other, less obvious truths and help us build a surer picture of the world. And he has a point. To disbelieve that 2 + 2 = 4, or think that all bachelors are married, would result in nonsense and contradiction. Their truth is therefore clear and distinct to me in a way that, for example, the contents of a dream or even a bodily sensation are not and to doubt them would be to call into question my ability to know anything at all.

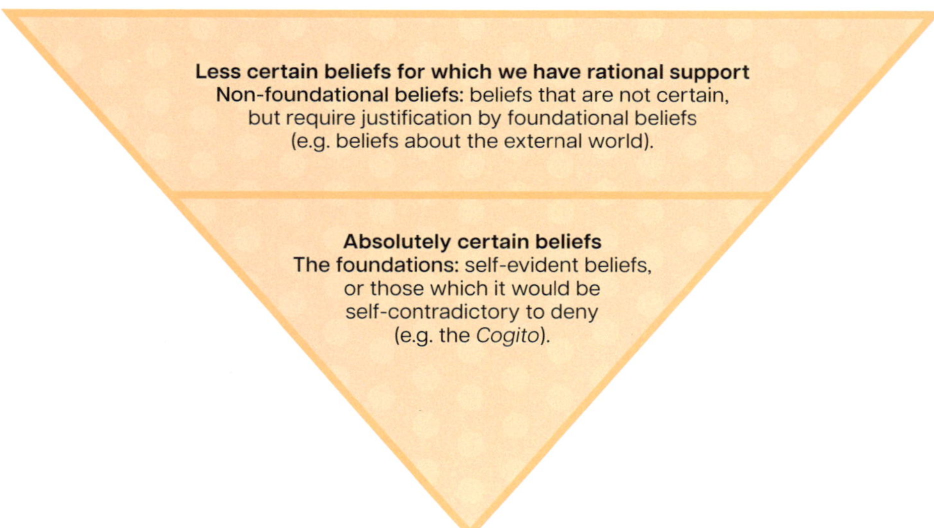


FOUNDATIONALISM

Foundationalism, then, as his approach has become known, seems like a sensible approach to knowledge. Instead of basing our beliefs on tradition, authority or what our parents or teachers tell us, we subject them to doubt and analysis, and try to trace how they are or aren't supported by more fundamental and certain truths. It's a bit like an upside-down pyramid.

However, there is a problem. What if he's wrong? What if the fact that an idea appears clear and distinct, or in some sense crazy to deny, is not in fact as conclusive as he thinks? After all, quantum physics implies that not all events have a fixed cause and the idea of the Big Bang may perhaps suggest that something can indeed come from 'nothing'. But the problem lies in this notion of a clear and distinct perception. How does he know that this experience is itself trustworthy? What if he's only fooling himself that this is the case?

It is at this point that Descartes pulls out his trump card: God. Being a benign deity, God would not create human beings in such a way that they could be deceived as to what they clearly and distinctly perceive to be true – would He? But what if God doesn't exist? What if, instead, there exists only some malevolent, all-powerful demon, intent on tormenting Descartes with falsehoods? But no, that's crazy talk! And besides, it is obvious from certain logically undeniable arguments that God does in fact exist, and that He is benign. So, Descartes concludes, he can in fact trust his clear and distinct perceptions regarding the truths of maths, logic and science.

If you've been following this argument closely, then you will have seen that Descartes has got himself into a bit of a muddle here. This has famously become known as the Cartesian circle. It's in that phrase: 'logically undeniable arguments' for the existence of God. For even if he did possess such arguments, the truth of them relies on his clearly and distinctly perceiving that such arguments are true. But...didn't he need God to justify that things being clearly and distinctly true was not an illusion? And now he's using his clear and distinct perception that God exists to guarantee...

That's right. It's circular. He needs God to guarantee trustworthy knowledge, and trustworthy knowledge to guarantee the existence of God.

If you are not a religious believer, then you might think this issue is irrelevant. But the point also applies to non-religious forms of foundationalism. For instance, let's say we replace God with a scientific argument, something along the lines of 'I know my rational perceptions are trustworthy, because if they weren't, then humans wouldn't have split the atom/landed on the Moon/mapped the human genome/[insert other impressive scientific feat here].' But this is also potentially circular, for despite all these achievements, it's still possible that any number of fundamental scientific beliefs are wrong, but they simply haven't affected the achievements in question. For instance, it's possible that certain physical laws may only apply at a certain place or time – now, in this galaxy – but perhaps not everywhere else. We just don't know.

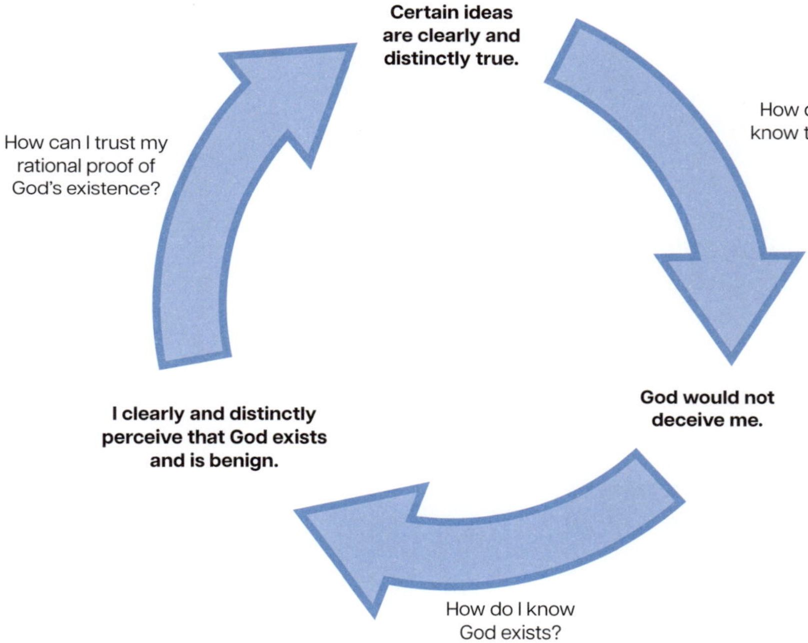


CARTESIAN CIRCLE

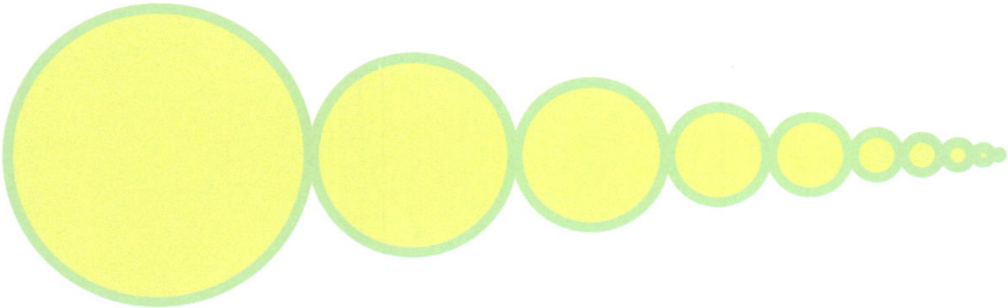

INFINITE REGRESS

So, we can replace God with Scientific method, or whatever you want, and you will still get a problem, because it's the *form* of the argument that's the issue. If we say, 'How do we know A is true?', then we can say, 'Because of B.' But how do we know that B is true? Either we support it in a way that eventually returns to A (it's circular), or else it goes on for ever (C, D, E, F, G...) in what's called an infinite regress.

So, it's foundationalism that is the problem, not your teachers or your parents, and not God.

# AM I LIVING IN A SIMULATION?

**You are out walking your dog in the woods one day when you see a squirrel. It runs away as your dog gives chase, leaving your hound barking in frustration as his fluffy quarry taunts him from the safety of the nearest tree. You laugh and walk on. A few minutes later, you see another squirrel, as does your dog, and the same thing happens again – in fact, so similarly, down to every precise detail, that it is almost like déjà vu, and you get the uncanny sense that it is the exact same event playing out again. Is it?**

The above scenario is similar to one depicted in the movie *The Matrix* (1999), where the main character, Neo, twice sees the same cat behaving in exactly the same manner. In the film, this is a sign that there is a glitch in the Matrix, where the robot overlords who have enslaved humanity are making changes to the virtual construct with which they have pacified their human slaves. During the movie, Neo has woken up to this fact, coming to realize that what he took for reality was no more than a computer-generated simulation.

Such an idea is not new. In Western philosophy, it goes back at least to Plato, who imagined that humanity existed chained up in a metaphorical dark cave of ignorance, where it struggled to interpret shadows on the wall. To grasp reality, it must free itself from these chains of prejudice and delusion and make its way up and out into the light of true understanding. In both Hindu and Buddhist philosophy, the world as most people experience it is known as *maya*, the illusory world of the senses that hides the ultimate nature of reality. We've seen earlier how Descartes used the 'malevolent demon' hypothesis and the possibly that life may be a dream to a similar purpose. Foreshadowing Descartes, Chinese philosopher Zhuangzi (*c.*369–*c.*286 BC) famously questioned how, on waking from a dream of being a butterfly, he could know whether he was not in fact a butterfly now dreaming that he was a Chinese philosopher. This type of scenario is commonly called the veil of perception argument, which goes something like this:

**1** Almost all of our information about the world comes through the senses.

**2** We have evidence that our senses sometimes mislead us.

**3** Therefore, it is possible that we are more misled than we like to think and are fundamentally mistaken about the true nature of the world.

It seems obvious that both (1) and (2) are true. So, does that mean that (3) is too? But what about science, you say? Well, yes, maths and physics do allow us to predict and control certain aspects of reality, but this does not guarantee that we have access to the ultimate nature of what is real. As Irish philosopher Bishop George Berkeley (1685–1753) argued, it's a logically coherent possibly that only ideas exist (a position known as *subjective idealism*), and that what we think of as physical things are just more permanent sorts of ideas or thoughts in the mind of God.

The veil of perception argument was recently updated by Swedish philosopher Nick Bostrom (1973–), who gave it a technological spin, which has become known as the simulation hypothesis. Computers are getting increasingly powerful. If they continue on this trajectory, there will eventually come a point where they can simulate the real world to the extent that no one can tell the difference between a virtual simulation and reality. If you were an anthropologist at such a future time, would you not make use of such technology to study your own ancestors' development, down to simulating the very thoughts and experiences of the virtual subjects themselves? And if so, who is to say that this has not already happened somewhere in the galaxy and that we are in fact living in such a simulation right this very second? In fact, if such simulations are possible, then they are statistically more likely to outnumber real worlds, just as our own world contains numerous games, simulations and models. Therefore, it's probable that we're now living in a simulation.

This may sound far-fetched. I've laid it out very briefly here, but in his 2003 paper 'Are You Living in a Computer Simulation?', Bostrom goes to pains to argue for the plausibility of each step in this argument. By the end of the paper, the hypothesis seems, if not likely, then at least more logically possible than you might at first think. But is it true? Bostrom argues that there are only three options:

**1** Either such advanced civilizations go extinct before developing simulation technology (as a result of climate catastrophe or nuclear holocaust, for example),

**2** They are simply not interested in running such simulations (perhaps they think it unethical, or whatever), or

**3** It's more likely than not that we are now living in a simulation.

It's an intriguing argument, one that has spread through popular culture and even developed a cult following in certain tech circles (you can even find people sincerely claiming to have spotted such glitches). But there are a lot of ifs that need to be the case for the hypothesis to have a good possibility of being true – *if* such a society survives, *if* it develops such technology, *if* it uses it to create sentient subjects.

Setting these points aside, the most fundamental question is whether we can apply the word simulation to a being's conscious experience of the world. Arguably, if such a future civilization succeeded in creating artificially sentient beings, then this is not a simulation at all (that is to say, an artificial replica of something), but conscious life itself, and those creators would therefore not be scientists, but gods. For even if the content of the experiences of such subjects were illusory, their consciousness itself would not be – which brings us back to Descartes and the *Cogito*.

Of course, the other problem is that, if the simulation is so perfect that we cannot tell it is a simulation, then the hypothesis can neither be proven nor disproven, in which case it is irrelevant. You may as well go about your life as usual.

Now, where did that damn dog go...

# WILL THE SUN RISE TOMORROW?

**Imagine you're a chicken. You wake up every day in a warm, well-kept barn. There is plenty of food and you are free to roam and roost and do whatever it is that chickens do. Life is good. Then, one day, the nice man who has kept you sheltered, fed and watered for as long as you can remember, comes along and wrings your neck.**

This is the problem of induction, as illustrated by English philosopher Bertrand Russell (1872–1970). Inductive reasoning is the process whereby we reach a probable conclusion about the unobserved based upon what we, or others, have so far observed. The Sun has risen every day of our lives and history records that it has always done so. Therefore, it will rise tomorrow. Induction may be contrasted with deduction, which is a form of argument where we reason from a set of statements to a certain conclusion, and has often been expressed as a three-step argument called a syllogism. The most common example of a syllogistic deductive argument is:

**PREMISE 1**

All humans are mortal.

**PREMISE 2**

Socrates is human.

**PREMISE 3**

Therefore, Socrates is mortal.

DEDUCTIVE SYLLOGISM

In deductive reasoning, as long as the first two statements (the premises) are true, and the argument is structured logically (valid), then the argument as a whole is sound, and the conclusion must therefore be certain. If we know that humans are mortal, and that Socrates is human, therefore we also know that someday he must die.

So, one requirement of a sound deductive argument is that its premises are true. But to give us useful information about the world, one of these premises usually has to state something that isn't obvious. And that's where induction comes in, for the premise 'all humans are mortal' is itself the result of inductive reasoning. This person died, then this person died, then this person died...throughout recorded history. Therefore, we reason from this set of observed facts to the general conclusion that all people who will ever live will eventually die.

But how do we know that the future will resemble the past? Perhaps Socrates is a new type of human that has evolved an immortality gene. How do we know that the finite amount of evidence we have is enough to guarantee that the rule will hold for everything else? Perhaps only a few humans are born immortal, so we are basing our conclusion on too limited a sample.

The depth of this problem was first highlighted by David Hume. Ancient philosophers such as Aristotle (384–322 BC) recognized that inductive conclusions were only probable but still considered them a good guide to discovering fundamental truths. However, Hume pointed out that because even our knowledge of those fundamental truths is mostly based on experience, this also makes it vulnerable to new discoveries. Like the chicken, we expect the future to resemble the past, and the universe to be uniform in nature, not realizing that we are merely being fattened up for slaughter. Yes, you say, but what about science? Surely, there are laws that accurately predict how the world will behave that involves maths and stuff. Isn't that a basis for certainty? Well, no, because even rigorous, mathematically defined laws are potentially subject to change in the light of new theories or discoveries.

For over 200 years, the laws of motion and gravitation discovered by English mathematician Sir Isaac Newton (1642–1727) were thought to apply throughout the universe, but now we know – thanks to German physicist Albert Einstein (1879–1955) – that they do not. Science abounds in such examples, and it is this process of replacing false theories with more accurate ones that defines scientific progress. However, no matter how much evidence we have, it means that inductive knowledge is never fully justified. Although, strictly speaking, there is such a thing as induction by complete enumeration: if I have a bag with six green apples in it and I examine each apple and discover it to be green, then I can conclude inductively that all apples in the bag are green. But this isn't hugely useful.

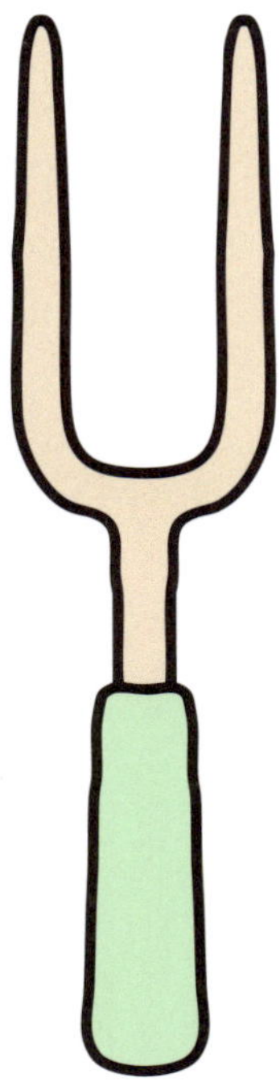


HUME'S FORK

Does all of this mean that we can't trust induction? Well, we can't really do without it. Without some inductive input, deductive reasoning would leave us simply restating things which are true by definition ('all bachelors are unmarried'), what Hume called relations of ideas. And since, Hume argued, our inductive knowledge is often based on long-established habit and custom (of seeing A cause B), then it would be foolish to dismiss it completely. In fact, it is so ingrained in our nature that we may be unable to do that. So, how do we proceed?

One solution is to distinguish between theoretical doubt (what Hume called extreme scepticism) and practical doubt (what he called mitigated scepticism). In practical terms, given everything we know, and what science gives us reason to expect, there is no good reason to doubt that the Sun will indeed rise tomorrow, so it is rational to believe that it will. However, in theoretical terms, we have no absolute guarantee that it will rise. At any minute, a rogue black hole might materialize and suck it into oblivion.

Of course, we can't live our lives as extreme sceptics – we must live practically, even if that means, from the theoretical point of view, that we are no different to the chicken.

# DID WE REALLY LAND ON THE MOON?

**It's been over fifty years since the Apollo 11 lunar module landed on the Moon. Despite all the documentary evidence, you can still find people who believe that it never happened, preferring instead the theory that it was filmed on a secret sound stage, or in the Nevada desert, with the footage slowed down to simulate the Moon's lower gravity. According to a C-SPAN/Ipsos opinion poll of 2019, at least 11 per cent of Americans were sceptical about the Moon landing. A survey in the UK by *Engineering and Technology* magazine in 2009 revealed that as many as 1 in 4 Britons shared these doubts.**

A cursory online search will find no shortage of scientifically robust rebuttals of the various conspiracy arguments but this seems to do nothing to change the views of the committed. You can even find an increasing percentage of people who believe the Earth is flat. Why is this? Why do people hang on to beliefs that appear to fly in the face of reason and evidence?

First, you should note here that I'm not saying all conspiracy theories are false. A theory is just a theory and should be taken on its own merits. A conspiracy is just a group of people meeting secretly to do something illegal or unethical, such as assassinating a politician, hiding corporate malfeasance or something else that they rightly think others might frown upon if they knew about it. By that definition, history is full of conspiracies that definitely did take place. But what most people think of when they hear the term 'conspiracy theory' is that the Moon landing was a hoax, that the assassination of JFK was an inside job or that a crashed flying saucer is kept in Area 51, a highly classified US Air Force facility in Nevada. The uncomfortable truth is that such things are hard to disprove – at least, to the point where it will convince all rational believers. And there is a very interesting reason for that.

The traditional idea is that science advances through testing and verifying theories. I have a theory (apples are subject to gravity), I test it (I drop an apple) and it is either confirmed or rejected (the apple falls or continues to float in the air). But having successfully dropped my apple, I can confirm that apples are indeed subject to gravity. Theory proved.

This view finds its ultimate philosophical expression in an approach known as verificationism (or logical positivism). This was popular in the first half of the 20th century and practised by such philosophers as A J Ayer (1910–89) and Rudolf Carnap (1891–1970). Drawing on the work of David Hume, verificationists argued that a statement is only meaningful if it can be verified by logic or experience. This caused problems for less quantitative fields, such as ethics (which verificationists reduced to emotional expressions) and theology (which they notoriously categorized as nonsense), but in terms of scientific progress, it provided a clear method of weighing up a theory: if a statement can't be tested or verified, then it isn't scientific.

However, it isn't that easy to conclusively verify something. For instance, to return to our apple-gravity theory, perhaps apples are only subject to gravity in certain places or under specific conditions. Maybe gravity doesn't apply to Braeburn apples or to Granny Smiths. We might also ask upon what basis we decide to believe that only verifiable statements are meaningful. As Austrian philosopher Karl Popper (1902–94) pointed out, that assumption does not in itself seem to be verifiable, and nor is it an obvious consequence of logic.

Popper would go on to suggest his own replacement for verificationism, which has become known as falsificationism. Rather than arguing that science advances by verifying theories, he proposed that it progresses through discarding false ones. If I have a hypothesis that Granny Smiths are immune to gravity, all I need to do is see my apple drop to the ground and my theory is refuted. I then move on to a new theory, concerning Golden Delicious or Pink Lady, perhaps.

The strength of Popper's approach is that, unlike verificationism, which relies on probabilistic *inductive reasoning*, our conclusion is certain, because it is *deductive*.

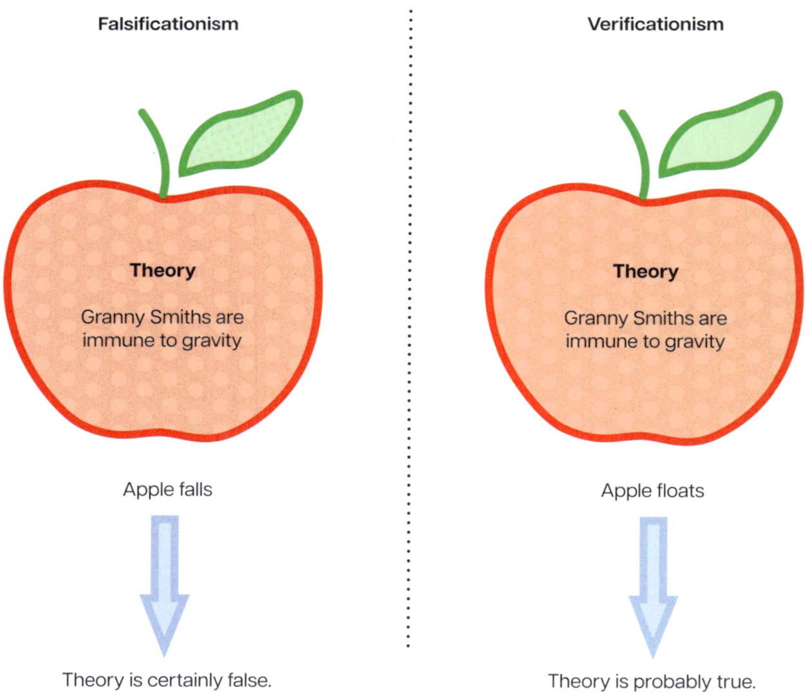


FALSIFICATIONISM VS VERIFICATIONISM

It should be noted here that Popper's approach did not propose that statements that could not be falsified were meaningless, but he did use it as a criterion for testing whether a theory was scientific. Famously, he argued that both the theories of Marx and Freud were pseudoscience because no evidence could be used to prove that their theories didn't work.

For instance, a Freudian analyst diagnoses his patient with a certain syndrome (the Oedipus complex, let's say), which the patient denies. So, the analyst accuses him of displaying resistance, a defence mechanism of the ego to protect it from painful truths hidden in the unconscious. But how are we to tell the difference between a patient who denies his diagnosis because he is resistant and one who rightly thinks that the analyst is simply wrong? In either case, the analyst can claim resistance and nothing can be done to disprove or, of course, prove, that diagnosis.

Here, then, we start to see a similarity with Moon landing conspiracy theorists who reject any alleged disproof of their viewpoint. Because they can always explain any disproof away ('the scientists and media are all in on it together', for example), then their view is ultimately unscientific. Nothing can falsify it.

There is some truth in this analysis, but it is not the comfortable win for rationality that it seems. That's because the history of science does not reveal the steady progress through the falsification of faulty hypotheses that Popper supposed. This was pointed out by American philosopher Thomas Kuhn (1922–96) in his 1962 book *The Structure of Scientific Revolutions*. Kuhn took as his example the overturn of the Ptolemaic world view, which stated that the Sun orbited the Earth (geocentrism), by the Copernican view that the Earth orbited the Sun (heliocentrism). Popper's theory would have predicted that, as more and more anomalies were found in the geocentric view, it would be abandoned for the more rational heliocentric one. However, Kuhn shows how this was not the case.

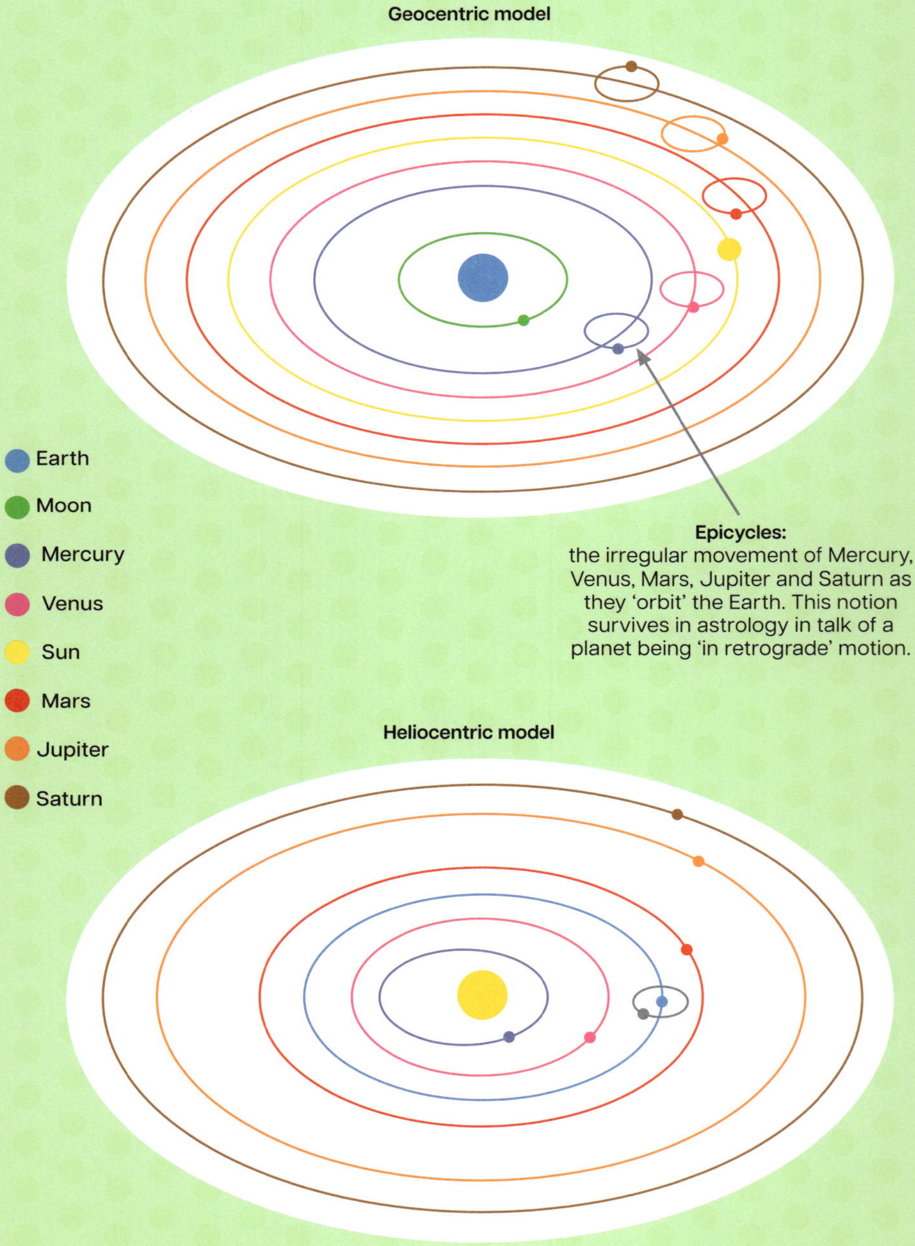


HELIOCENTRISM VS GEOCENTRISM

What actually happened was that for a long time the geocentric model was continually adapted to accommodate anomalies. Eventually, there was a sort of mass desertion, what Kuhn termed a paradigm shift, as scientists were persuaded over to the new model. Kuhn argues that what motivates such paradigm shifts is not just the weight of evidence against the prevailing theory, but a number of social, psychological and practical factors. For instance, the new model provided fresh and exciting avenues for research and coincided with a growing political and religious dissatisfaction with the status quo, with which the old Ptolemaic system was associated.

Therefore, if Kuhn is right, scientific progress is not the purely rational process that Popper and others had claimed. This is not to say that Kuhn was arguing that paradigm shifts were irrational, but merely that proof and conformity to reason were not the only factors. And once a paradigm had become established, its fundamental tenets were not questioned, but merely accepted, allowing working scientists to conduct what Kuhn called normal science.

This brings science uncomfortably closer to conspiracy theories than we'd like to think, and more akin in some ways to the belief systems we find in religion. In both cases, faced with what psychologists call cognitive dissonance (the simultaneous holding of incompatible views), both the conspiracist and the scientist will likely reject the incompatible evidence rather than jettisoning their whole theory. A scientist might put the anomaly down to faulty equipment; the conspiracist to faked evidence. Ultimately, both are playing a similar game.

# IS THE MAJORITY ALWAYS RIGHT?

**You are called for jury duty. It is a complex murder case with many twists and turns, and as the jury retires to reach a verdict, you are surprised to find that you are the only person who thinks the accused is innocent. Are you wrong? As you discuss the case with your fellow jurors, you find yourself becoming even more certain that there is not enough evidence to convict, although you are unable to convince anyone else of this.**

Secretly, you suspect that many of the other members of the jury harbour an irrational bias against the defendant. It's deadlock. However, because it has been a complex case, and a long and expensive trial, the judge reluctantly decides to accept a majority verdict (as opposed to a unanimous one). You are outvoted. The accused is found guilty.

Has justice been served? There are many situations in social and political life where opinions are divided and the matter is carried by a majority vote. Everything from who runs the country to what to watch on TV can come down to a simple question of numbers. You might feel this is unfair, especially if you're in the minority and it concerns your right to behave or express yourself in ways of your own choosing. The result is what English philosopher John Stuart Mill (1806–73) called the tyranny of the majority.

Should smoking, skateboarding or street magic be banned in public spaces? Well, if the majority are non-smoking non-skateboarders who aren't keen on card tricks, then fans of these activities will simply have to give way. However, as the jury example shows, majority opinion doesn't just decide on matters of taste and personal preference but can also be used as a justification for claiming to know the truth.

Plato argued that to know something we need to have a justified, true belief. In other words, we believe something that we have justifications to think is true, and it turns out to be so. I am justified in believing that I locked the front door after leaving the house this morning, because I have a mental image of having rattled the door handle before walking away. When I get home, the door is indeed locked, and so I was right: I knew that I locked the door.

However, very few things in life are absolutely certain, and sometimes we claim to know things that may turn out to be otherwise. Perhaps, even though you returned home to find the door locked, unknown to you, your partner had come home earlier in the day and discovered it unlocked, and then locked it when they went out again, all before you got home. The mental image you had of rattling the handle was from yesterday, what you thought was true was only true because of a chance sequence of events, and you did not have sufficient justification for your belief.

As this example shows, while we might aim for it, the truth is frequently beyond our reach. For this reason, it is perhaps better to focus on justification. To return to the jury example, we know the facts of the case, have heard the evidence of witnesses and the arguments of the defence and the prosecution, and from these we must decide if our belief that the accused is innocent or guilty has sufficient justification.

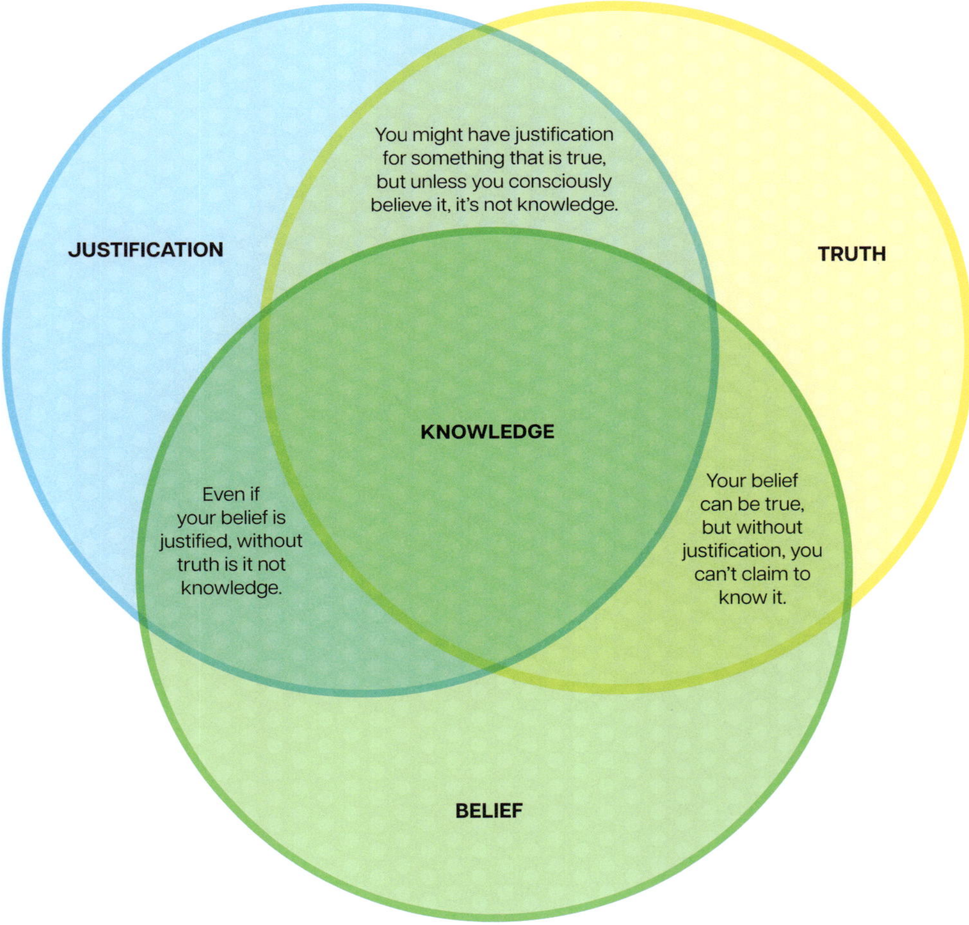


PLATO'S TRIPARTITE DEFINITION OF KNOWLEDGE

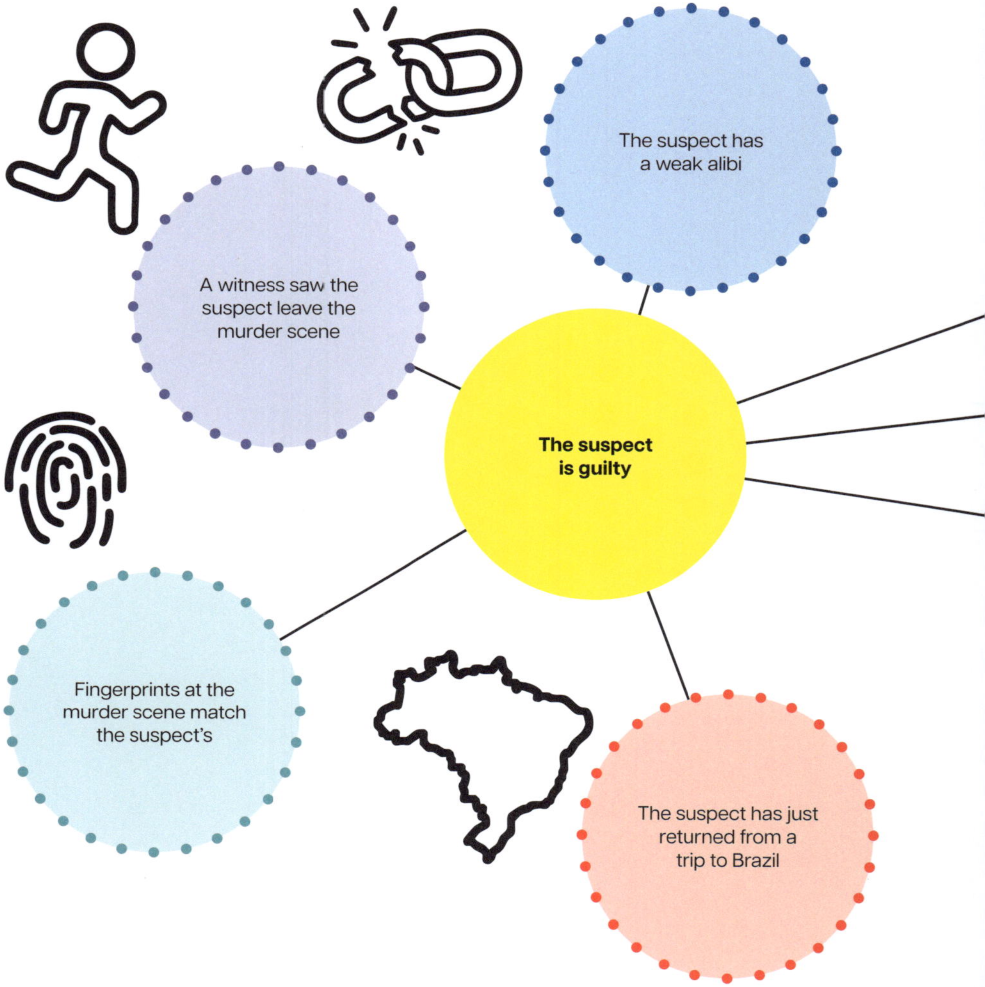


Taken together, the various beliefs make a coherent case for the suspect's guilt – but is that the only interpretation that fits the facts?

A COHERENT CASE?

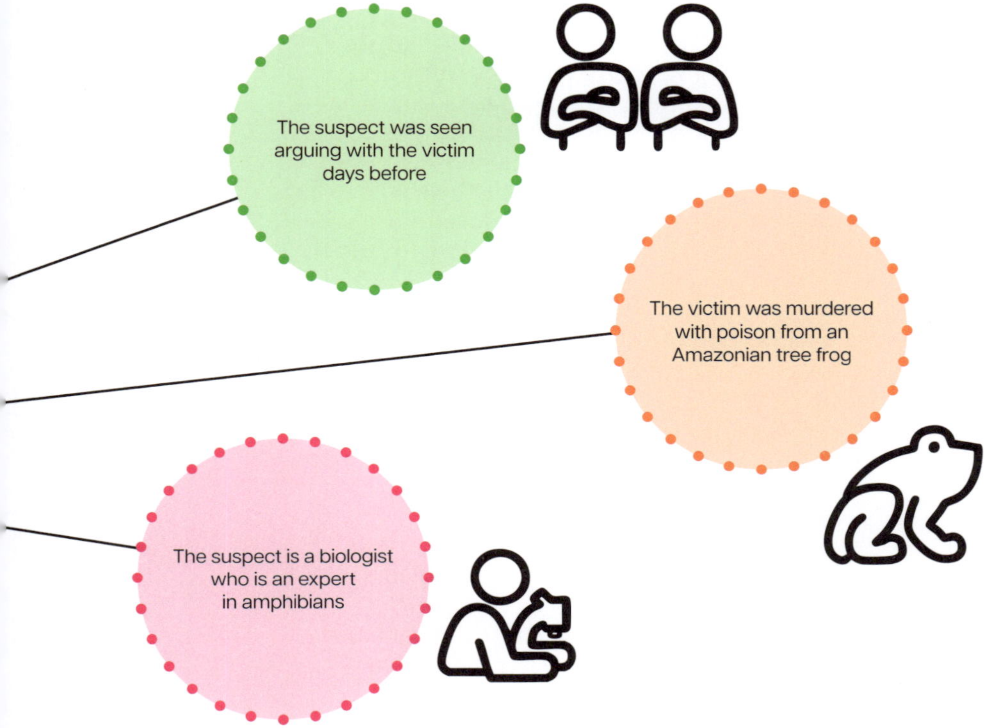


In other words, it's a matter of interpretation of the facts and trust in the process. We must believe that the judge was fair, the police did not plant evidence and the witnesses were reliable and did not lie. But as to what actually happened, well, in the absence of film footage of the murder, that might be something that we never have direct access to (and even film footage can be faked these days). If so, then perhaps the existence of some external, absolute, objective truth is less important than having sufficient justification for it. But how much is ever enough?

We've seen that, for Descartes, what makes most of our everyday beliefs justified is the fact that they are built on an ultimate foundation of certainty. But since we've also seen that this has its own problems, some philosophers have argued that what matters more is the internal coherence of our beliefs. Our claims to knowledge are more likely to be rational or justified if they fit with other beliefs we hold to be true. I am excited to read a new book by my favourite author, I have read reviews that confirm it's great, and so my belief that it is worth buying is made more coherent.

As such, while foundationalism sees justification in terms of an upside-down pyramid, with the most certain beliefs at the bottom, coherentism sees things more in terms of a spider's web, where each strand of belief is mutually supported by others. It is therefore the overall structure, the coherence of the whole, that matters most.

It's worth noting here that coherence itself doesn't make a belief true – it only makes your belief more justified. A wrong theory can be coherent. In the jury example, the majority's irrational bias makes the defendant's guilt a more coherent story for them than his innocence. And the fact

that you think the complete opposite suggests that there can be more than one coherent story, each explaining and interpreting the facts in a different way. It's even possible that you have your own irrational biases in favour of the defendant.

So, we shouldn't confuse truth with evidence for truth (or justification). If I can amass more evidence for my beliefs by making them more coherent, that doesn't necessarily make them more likely to be true, because coherence can be achieved through, for example, confirmation of my own biases.

What does all this mean for truth and our ability to know it? If all we have is justification and the truth itself is perhaps inaccessible, does it mean that, effectively, it doesn't exist?

Well, philosophically at least, let's just say that the jury is still out.

# DO BABIES KNOW ANYTHING?

**Babies are enigmatic – when they're not crying, pooping or sleeping, that is – and there's a certain sort of knowing look that can come over them that almost suggests a sort of wisdom. But do babies actually know anything?**

This is an old debate and concerns the possibility of innate knowledge. Is the mind a blank sheet of paper from birth, or does it come pre-loaded with basic instructions, precepts and rules? As we've seen, Plato supported the latter idea (see page 114), believing that the soul reincarnates from life to life, bringing with it knowledge from a pre-birth existence. In his philosophical dialogues the *Phaedo* and the *Meno*, Plato argues that the only reason we know certain truths in this life is because we were born with this innate knowledge. Therefore, we must have existed in a disembodied state before birth where we had first-hand aquaintance with such eternal truths. As such, all learning is just recollection of this pre-birth knowledge.

In the *Meno*, Socrates interrogates an uneducated enslaved boy as to his mathematical knowledge. While the boy has had no formal education, Socrates attempts to show how he nonetheless possesses the innate seeds of mathematical principles and all that is required to bring them out is to ask the right questions. It's debatable here whether these questions are, as Socrates claims, merely a catalyst for drawing out the boy's own knowledge (what has become known as the Socratic method), or, as Plato's critics argue, merely a way of leading the witness, to borrow a legal term, by suggesting the answer.

Whatever the case, the idea of innate knowledge is generally associated with a position known as rationalism. Plato shares this view with Descartes, who, in addition to mathematical truths, argued that we also possess the innate knowledge of God's existence.

Again, the implication here is not that babies have an intellectual conception of rational arguments – Descartes did not think that an infant emerges from the womb quoting the *Cogito* – but rather that there are certain types of knowledge that could not be arrived at through experience. These are, in this sense, *a priori*, to use the Latin term (that is, known 'prior to' or independent of experience). We need experience to bring this knowledge out, but the knowledge is not reliant upon experience. The problem with this is how we prove it. As we saw with the Cartesian circle, such foundational truths would seem to either require no proof (they are obvious dogmas), or we are left running in logical circles trying to support them.

We may contrast this approach with that of empiricism. For thinkers such as Aristotle, John Locke and David Hume, the mind was a blank slate waiting to be written upon by experience. For Hume, all experience is either an impression or an idea. Impressions can be subdivided into sensations or reflections. Sensation concerns the vivid, immediate experience of something with the external senses (sight, touch, taste, hearing and smell). Reflection describes the internal stirring of emotion or desire experienced when reflecting on an idea.

For example, the idea of a spider might cause fear, while seeing a picture of a loved one might inspire a desire to see them. In contrast to both these forms of impression, ideas involve recalling, imagining or thinking about these sensations and reflections in a less vivid form after the fact. As Hume puts it, 'the most lively thought is still inferior to the dullest sensation' (*An Enquiry Concerning Human Understanding*, 1748).

So, I see or smell a rose (sensation) and later on I can remember that experience (idea) and, by reflecting on it, be filled with a feeling of pleasure (reflection). Put simply, we're either perceiving things with our external senses (sensation), responding to memories, imagination or thoughts via emotion or desire (reflection) or else thinking about what we've sensed or felt or thought (ideas).

So how do we build up our knowledge from this? Let's take an example. I have seen a ripe tomato and I have seen a red rose. When I recall my mental images of these two items, I notice they are visually similar and I can abstract the idea of redness. I can do the same with the roundness of a ball and the full Moon, or by contrasting the texture of human skin (smooth) and that of tree bark (rough). Furthermore, by

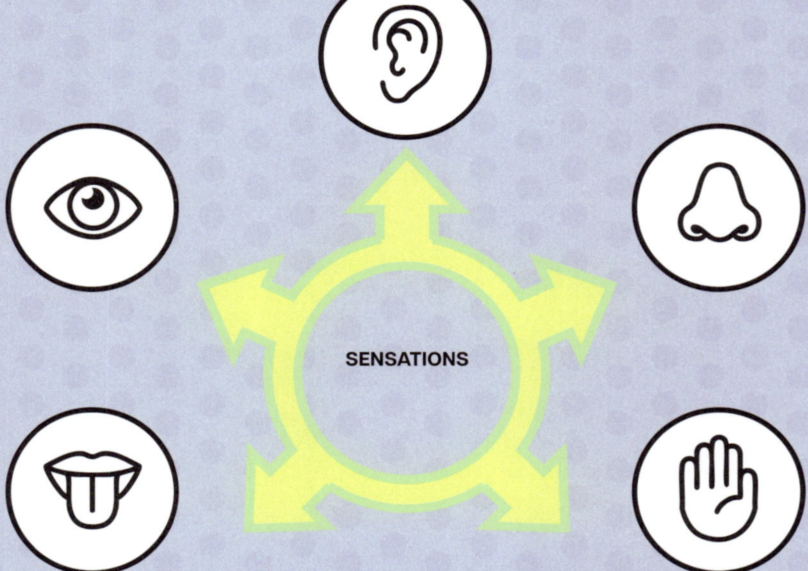


Information received through the external senses

HUME'S MODEL OF THE MIND

Emotions or desires from internal reflection

HUME'S MODEL OF THE MIND

comparing, contrasting and combining these basic notions (redness, roundness, smoothness), I can come up with more complex and sophisticated ideas. So, even a concept as lofty as that of God – which rationalists such as Descartes agued were so unique that they could not come from experience, and so must be innate – may be thought of as being made up of simpler, more humble ideas. If strength, knowledge and compassion are expanded and amplified to a much greater degree, we get omnipotence, omniscience and divine love.

So, to paraphrase Aristotle, there's nothing in your mind that wasn't first put there by your senses, or been derived from them, and all our knowledge is therefore *a posteriori* (acquired after experience and dependent upon it). For Hume, innate ideas don't exist, because they don't need to.

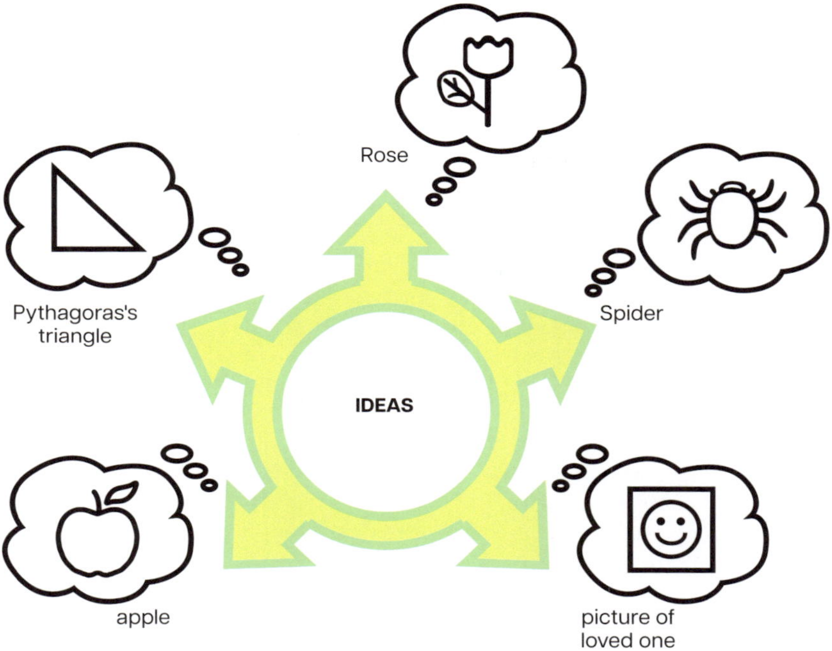


HUME'S MODEL OF THE MIND

But this view too has its problems. As we've seen, Hume argued that our belief in cause and effect arises from our long-established habit of associating certain outcomes with particular actions. However, as he also acknowledged, this does leave a gap in which the sceptic can point out that the effect of a cause is not guaranteed. It's not against logic to argue that tomorrow the Sun may not rise or that apples won't fall. Once we acknowledge that all of our knowledge is based on sense experience, then it opens us up to the possibility that we can have no certain knowledge of the cause of the physical world itself – or its existence. And we're back with Descartes's deceiving dreams and demons.

If rationalism consists of unprovable dogmas and empiricism opens us up to extreme doubt, is there a way in which we can retain the best of both approaches? The German philosopher Immanuel Kant (1724–1804) thought so, arguing that the fundamental truths that the rationalists sought to prove were not in fact innate ideas, and nor, as the empiricists argued, were they derived wholly from experience. Rather, they were pure concepts of the understanding, built into experience itself.

This is a subtle theory, but it is best understood in terms of the idea that the mind itself needs an in-built form and structure to make sense of the world. So, we aren't born with a fully fledged rational idea of cause and effect, but rather we are mentally hardwired to see the world in terms of it. It's like wearing a pair of glasses – without them, the world is a blurry mess; with them, everything is brought into sharp focus. Now imagine that the glasses are sort of mental intuitions we possess that shape the world for us, ordering it into objects and the way they behave and interact with each other. We don't conclude that objects simply pop into existence out of nowhere for no reason. Instead, we unconsciously presume that, unless something affects them, objects have a certain nature and ways of behaving.

It's a neat solution. But is Kant right? Well, there's no real way to tell. The only way to know what the world looks like if you're not wearing glasses is to take them off, and we can't do that with our minds. Maybe, somewhere in the universe, there are creatures who experience the world differently. Such a possibility is the premise of Ted Chiang's novella *Story of Your Life*, filmed as *Arrival* (2016), where first contact is made with the Heptapods, extraterrestrials who experience temporal events not sequentially in terms of past, present and future, but all at once. So maybe, if such beings exist and we were to meet them, we could compare our mental glasses.

# DO I REALLY KNOW WHAT I THINK I KNOW?

**So far, we have considered several sceptical arguments concerning our ability to know fundamental things. It is possible that life may be a dream, or that we may be systematically deceived as to the true nature of reality. It may be that our knowledge has no secure foundation and that there is no guarantee that even what seem like justifiable beliefs are true. In fact, it may be that truth itself doesn't exist and is a mere social convention with no real basis. It could also be that what we think of as scientific laws don't apply in other times and places, and there is no way to prove that the future will resemble the past.**

You will be heartened to know that this does not exhaust the sorts of problems that philosophy raises. As you go about your day putting your trust in reality, be assured that you have absolutely no right to do so! Or at least, that is what a certain type of philosopher would want you to think. Such a person is known as a radical sceptic, and they have existed from the very beginning of philosophy. One of the first, most prominent sceptics was the Greek philosopher Protagoras (*c*.490–*c*.420 BC), who famously stated that 'Man is the measure of all things.' We're not completely sure what he meant by this, as only selected fragments of his writings survive in the works of other philosophers, but he would seem to be implying a sort of relativism or anthropocentrism. That is, the idea that humanity remakes the world in its own distorted image.

As we've seen, sceptical arguments can play an important role in positive philosophical views. Descartes used them as part of his method of doubt as he tried to isolate the types of ideas that give us absolute certainty. Hume used them to show that our knowledge of the world was more limited than previous philosophers had assumed and that if we are to find reliable knowledge, we must understand the role that experience plays.

In contrast, radical sceptics argue that there is nothing we can know, nothing we can be certain about, no such thing as truth, no fixed moral values or codes, and basically that everything is up for grabs. The sceptical philosopher Gorgias (*c*.483–375 BC), who was a contemporary of Socrates, took a fairly hard line on the matter. He said that nothing exists, and even if it did, it would be impossible to know or say anything about it. Well, I think that covers everything!

Plato spends many of his dialogues arguing against the cynical positions held by sceptical philosophers known as sophists. These men were commonly employed as freelance tutors to teach oratory and rhetoric, or how to reason, speak and write well. While teaching these skills, a sophist might adopt a sceptical position merely to win their case. Protagoras himself, credited by Plato as being one of the chief sophists, was reputed to have said that 'in every question there were two sides to the argument exactly opposite to one another' (Diogenes Laertius, *The Lives and Opinions of Eminent Philosophers*, 2nd century AD). It was this flexibility and disrespect for the truth that enraged Plato. Philosophy should not be a tool for getting whatever you want, but a guide to truth and how to live.

It is hard to argue against such radical scepticism. For any positive statement, the sceptic can always simply counter, 'But how do you know that?' or 'How can you be sure that you aren't deceived or mistaken?' They are like an annoying, precocious child, forever asking 'Why?'

It's possible that the sceptic's determination to question absolutely everything opens a way to combat them. We might ask whether questioning everything leads to incoherence. If I say, 'No statements are true', then what about that statement itself? Scepticism is a standpoint like any other philosophical position and, as such, must rely on certain things being true (such as that statements are either true or false) in order to question others. Of course, the sceptic need not assert this; they might only argue that we don't know which statements are true, which is a subtler position. But in doing this, they must also admit that some things must be true. Gorgias's claim that no truths exist or can be communicated would seem to rule out our taking his own pronouncements seriously.

There is also the question of whether radical scepticism asks too high a standard of knowledge. Even if many of our beliefs lack absolute justification, that doesn't mean they are not practically useful and can serve as true in various contexts. This is a position known as pragmatism, which is associated with American philosophers such as William James (1842–1910) and C S Pierce (1839–1914).

Pragmatism suggests that in giving up the Cartesian notion of absolute certainty, we can still have knowledge in an everyday practical sense. For pragmatists, this involves adopting a fallibilist attitude, which supports the idea that, if we cannot have absolute certainty or justification for our beliefs, then it is best to settle for a lesser form of knowledge. This doesn't mean that we can't know anything, but merely that, while claiming knowledge, it's not absolutely out of the question that we can be mistaken. In this way of thinking, beliefs are tools that we use to navigate reality, and some tools turn out to be more useful than others.

This takes a lot of the sting out of the radical sceptic's arguments, for we can still contrast knowledge with things we are less sure about. By adopting this *gradualist* approach that there are grades of knowledge and certainty, we can put more trust in beliefs for which we have a good amount of evidence, over those for which we do not.

This may not be the knock-out refutation of scepticism that philosophers would hope for, but it at least gives us a means to go about our daily lives with the assurance that some things are truer than others. Probably.

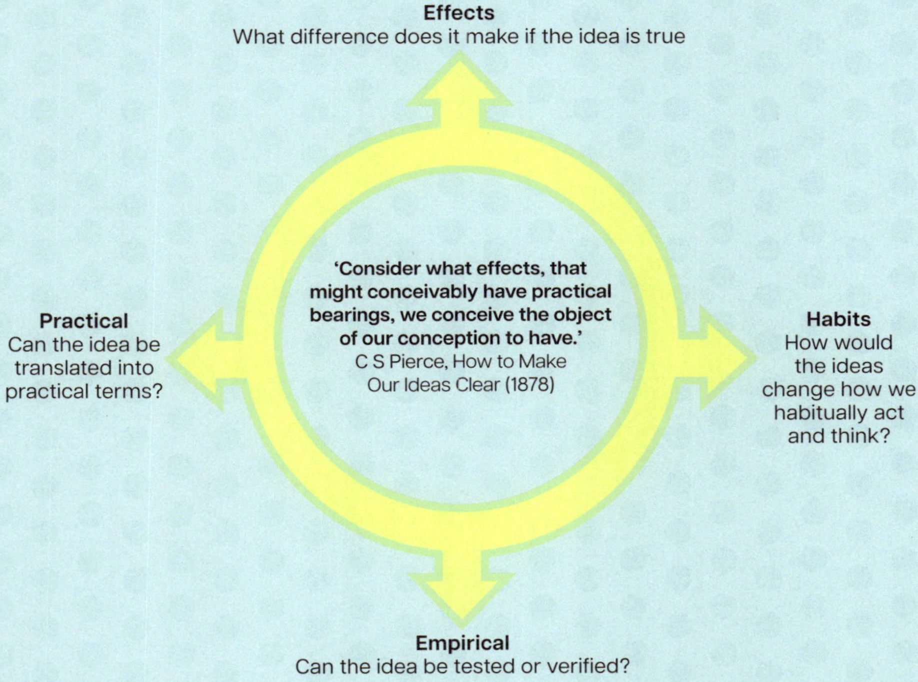


For pragmatists such as Pierce and James, the value of an idea lay not so much in its 'truth', but in how it might affect our beliefs and actions.

THE PRAGMATIST ATTITUDE TO TRUTH

# LIVING THE GOOD LIFE

## ETHICS

# 3.

# IS HONESTY REALLY THE BEST POLICY?

**We've all told white lies. We've probably all told some great stinking black ones, too. Some politicians and CEOs even turn it into an art form. Despite how common lying is, it is also universally condemned, and to call someone a liar still has a sting to it. But is it always immoral?**

If you think so, then you probably think of ethics as a set of rules that we are obliged to follow. This is known as the deontological view, which sees moral choices as being duty-based. A religious deontologist might point to the Ten Commandments, or Christ's maxim that, 'whatever you desire for men to do to you, you shall also do to them' (Matthew 7:12). However, Immanuel Kant argued that our moral duties have a rational basis.

Something that you *must* do is an imperative, or a command. Now, imagine that you want to lose weight. If you want to do so, then you must diet, visit the gym or shell out for one of those new weight-loss drugs that all the celebrities are taking. However, in most cases, we don't have a duty to diet or exercise – it's just something we should do if we want the result. As such, it is merely a hypothetical imperative, conditional upon our own desires.

Now, contrast that with telling lies. You don't tell the truth because you want something else – or at least, to do so would not be a moral reason. I might lie to save myself from embarrassment or to evade legal consequences. But neither of these are moral motives. In fact, the only true moral motive is to do the right thing for its own sake. That is, to act only out of moral duty – what Kant called the good will. But why?

Kant argued that in addition to hypothetical imperatives (things we must do if we want something else), there are also categorical ones (things we must do because not to do so would be irrational). So, 'do not lie' is, in Kant's eyes, an example of a categorical imperative and we have an unconditional duty to obey it.

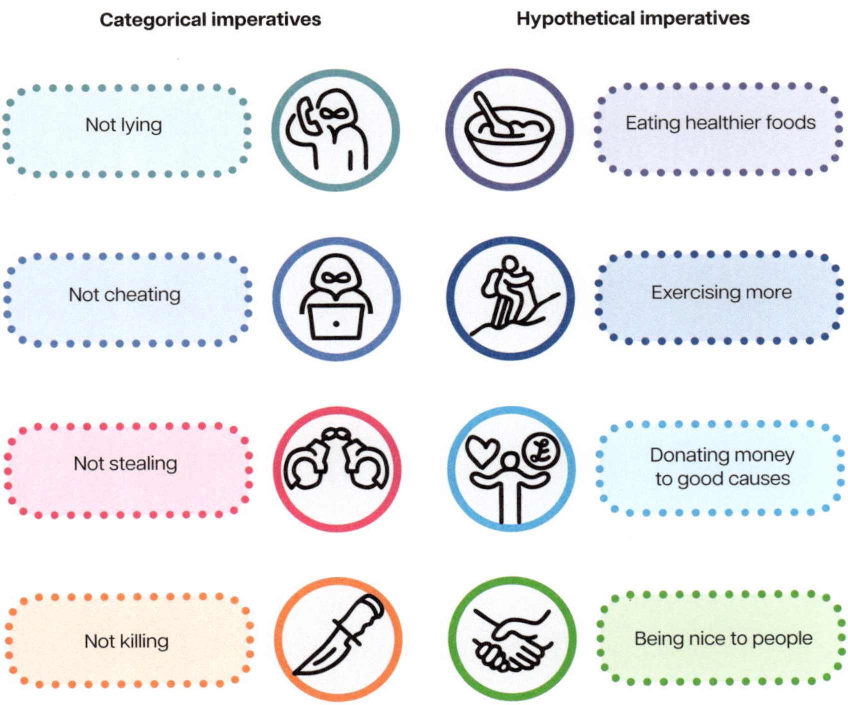


THE DIFFERENCE BETWEEN IMPERATIVES

Kant gave various explanations of why breaking such rules is problematic, which develop from each other and overlap, but are usually seen in terms of three formulations.

First, imagine if we all lied whenever we felt like it and no one considered it a bad thing. There would be no such thing as the truth, society would break down and all lawyers would be out of a job because no contracts would be enforceable. This demonstrates that if we were to make a universal law out of 'it's fine to lie', it simply wouldn't work. Moral rules are therefore universalisable: they need to apply everywhere and to everyone.

The second formulation argued that we should never treat others as a means to our own ends, but as ends in themselves. I shouldn't be friends with you because you're a generous person who owns a van and can help me move stuff when I need it. You might have your own plans that you've kindly put on hold to help me out. In acknowledging that, I respect your autonomy – your right to pursue your own rational goals – and my friendship should honour and facilitate that.

Combining the above principles, the third formulation suggests that we may each think of ourselves as existing in a moral community, or a kingdom of ends, as Kant termed it. This community is regulated by universalizable laws that apply to everyone no matter who they are or what situation they find themselves in. It is a community where the citizens treat each other as ends in themselves, never solely as means to their own ends, thus respecting each other's autonomy.

Now, if you think lying is sometimes forgivable, then you might feel Kant is being a bit inflexible here – and you wouldn't be the first. What if you lie to protect someone's feelings or even their safety? Should you be bound to tell your partner that, sorry, that dress does make them look fat or the pattern on their favourite tie reminds you of vomit? Or, more seriously, should you be obliged to reveal the location of your valuables to a thief who's threatening you at gunpoint?

A consequentialist theory, such as utilitarianism (which we'll cover shortly, see page 96), would have no problem dealing with any of these scenarios, because what is moral is judged according to the consequences of an action and if they're bad, then we are free not to choose them. But deontologists must account for situations where rigidly following a rule leads to worse outcomes or even into conflict with other rules. For instance, we might think it a rule to protect the lives of others if we can, but also that to take another's life is morally wrong. But what happens if we can't do the former without also doing the latter? Or what if someone intent on murder asked us the location of an innocent friend?

Of course, there are ways around this. The lying-to-a-murderer objection, as it's sometimes known, was raised in Kant's lifetime by Swiss thinker Henri-Benjamin Constant (1767–1830). Kant responded by stating that it would indeed be immoral to lie, even to protect the innocent friend. However, there is nothing in the categorical imperative that says that we cannot refuse to answer. Which seems like a neat solution. You can imagine that someone who has taken a vow of honesty might find all sorts of creative ways not to lie without also telling the truth. However, this is an evasive solution, and all that's needed to raise the problem again is to picture a situation where a response is required. What if the murderer threatened to shoot a family member unless you told them where your friend was? What if he tortured you? Sometimes, you just have to lie.

# CAN I DO WRONG TO DO RIGHT?

**You grow up in the perfect city. There's no conflict, no racial strife, no poverty or malnutrition. There's no need for rulers, priests or lawyers, as everyone lives in peace and happiness. Then, on your 18th birthday, you are taken to a hidden room where the truth behind all of this perfection is revealed to you. To keep it going, a single innocent child, who is perpetually imprisoned and abused, must suffer. You have a choice: to leave this civic paradise for ever or to accept this injustice as the price of everyone else's happiness. What do you choose?**

The basis of this thought experiment is Ursula Le Guin's 1973 short story 'The Ones Who Walk Away from Omelas', which is a perfect illustration of the type of dilemma that faces any form of ethical consequentialism. Unlike deontologists such as Immanuel Kant, consequentialists argue that what makes a moral decision right is the outcome. If an action can be justified by its consequences, then it is morally correct.

The most popular form of consequentialism is utilitarianism, which originated with English philosopher Jeremy Bentham (1748–1832). As human beings, Bentham argued, we are fundamentally driven by a desire for pleasure, which is the basis of what we think of as happiness. All of our actions, therefore, naturally seek to maximize our own happiness and minimize our suffering. But since what makes one person happy might make another miserable, the role of morality is

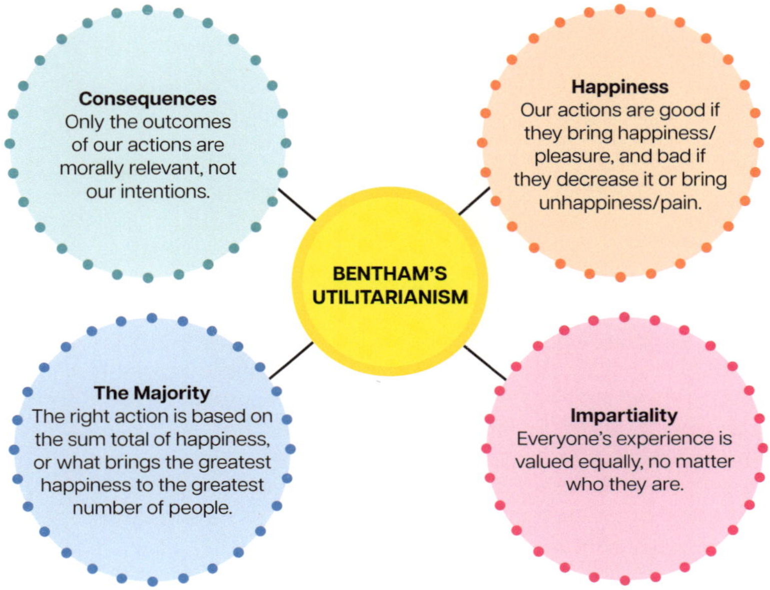


BENTHAM'S UTILITARIANISM

to promote only those things that are rationally justifiable. This makes utilitarianism a democratic philosophy, as it states that we should adopt only those actions that maximize the happiness of the greatest number of people. If it makes the majority happy that murderers face the death penalty, then that's the moral thing to do, because it outweighs the suffering of the executed man and the grief of his family.

Bentham's approach is known as act utilitarianism, because it focuses on the justification of individual actions. This means that an act utilitarian is not bound by any moral rules or edicts and can justifiably change their position if the situation dictates it, just as the majority in a democratic system can change whom it supports from election to election.

This flexibility is also responsible for some of utilitarianism's more controversial aspects. In the example of the city of Omelas, an act utilitarian might justify the suffering of the child because it promotes more overall happiness. What's one person's suffering weighed against a whole city's peaceful prosperity? This might seem repugnant to you, but it illustrates the sort of decisions that politicians and lawmakers take every day. You can't please everyone and if a good thing requires a minority to be unhappy, then what can you do? What Le Guin's story highlights, however, is the fundamental injustice of this approach. If utilitarianism leads to something as unconscionable as the intentional suffering of an innocent child, then surely there's something wrong with utilitarianism.

In light of such criticisms, other philosophers have attempted to soften Bentham's approach. For instance, John Stuart Mill advocated for rule utilitarianism, which attempted to address some of the issues that spring from judging the rightness of moral decisions based on a single action. Mill proposed that we think of whether certain moral rules create more happiness or less. Maybe, in a certain instance or a particular time, an action could produce more happiness, but if over time and in different places, it would not, then it isn't justified. Long prison sentences for people who steal parking spots might find mass support in a city hosting a World Cup final, but where passions are cooler and the context different, the majority would not favour this approach.

Mill's other innovation was to broaden Bentham's notion of happiness. Bentham was notoriously indifferent to different forms of pleasure, famously stating that: 'Prejudice apart, the game of push-pin [a trivial parlour game of the time] is of equal value with the arts and sciences of music and poetry' (The Rationale of Reward, 1825). His approach is therefore called hedonistic utilitarianism, in that it involves a notion of happiness that is primarily based upon pleasure in basic quantitative terms.

Instead, Mill argued that there are in fact higher and lower forms of pleasure, and that we should mostly prefer the latter. As he put it: 'It is better to be a human being dissatisfied than a pig satisfied; better to be Socrates dissatisfied than a fool satisfied' (*Utilitarianism*, 1861). In

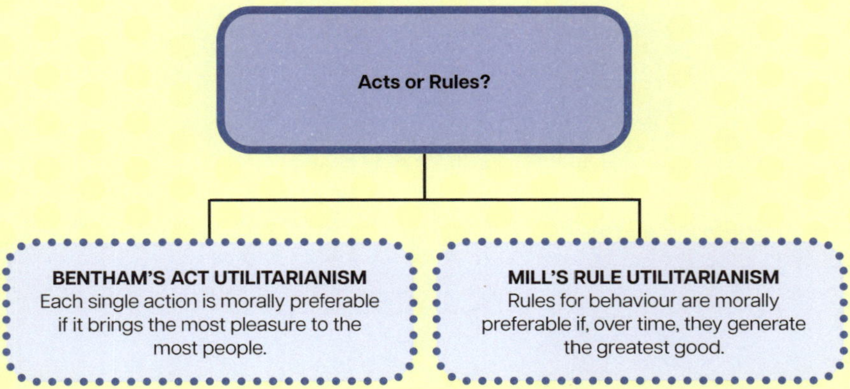


BENTHAM AND MILL'S UTILITARIANISM

other words, the complex mental life, which might include its emotional ups and downs, is still preferable overall to an existence that just satisfies our basest instincts. That's because the lesser amount of pleasure that thinking might involve is greater in quality than mere bodily pleasures.

Rule utilitarianism does solve some of the issues faced by its act-based alternative, but it is still open to criticism. For instance, in promoting a rule-based system of morality, it risks returning to a form of deontology that loses the positive flexibility that act utilitarianism allows. It loses the ability to say: 'This action is best because, regardless of the rules, it simply creates the most happiness and least suffering in this situation'.

Is the suffering of one person justified by the happiness it brings to the majority?

UTILITARIANISM AND THE PROBLEM OF JUSTICE

It is this approach, for instance, that allowed Bentham to advocate for animal rights. He argued these should be judged purely in terms of happiness and suffering, of which animals are as capable as humans, despite the 'rules' laid down by theologians or the theories of philosophers such as Descartes. We might also argue that Mill's concept of higher pleasures is elitist. What does it matter if I prefer boxing to Beethoven or hamburgers to haute cuisine?

Other philosophers have rejected the pleasure basis completely. For instance, English philosopher R M Hare (1919–2002) advocated for preference utilitarianism, which states that it is moral to maximize a person's preferences or interests. I might prefer to be vegetarian, which arguably doesn't impinge on anyone else's interests, so I should be allowed to pursue the actions that promote it. However, such preferences might be subjective and could conflict with the preferences of others. What happens when one person loves lawn bowling but someone else wants to turn the bowling green into an ice-skating rink? How can conflict between such personal preferences be resolved?

Without a common measure (for example, happiness or pleasure), there is arguably no solution. From this point of view, the decision of whether you do or don't walk away from Omelas would seem to come down to personal inclination.

# CAN I HAVE TOO MUCH OF A GOOD THING?

**Virtual reality has come a long way. In the foreseeable future, it might one day be impossible to tell virtual experiences from real ones. This illusion could be further increased through paraphernalia that stimulate sensations of touch or smell. As such, there might even come a point where some people prefer to live in the virtual world, experiencing a succession of purely pleasant experiences devoid of pain and disappointment. Does this appeal to you?**

This scenario is taken from *Anarchy, State, and Utopia* (1974), by American philosopher Robert Nozick (1938–2002). He argues that, faced with such a choice, there would be good reasons not to plug in to such an experience machine. Among them, Nozick cites the desire to actually achieve things and not simply have the pleasurable consequences of having done them.

So, you want to be a successful writer, footballer or virtuoso cellist. The experience isn't just about the pleasure of having lots of fans, or trophies or standing ovations. It's also about the process of striving and achievement, of learning and acquiring skill, which may not always be a purely pleasant one. The hard practice, the frustrations and setbacks, can't be divorced from the lives of a successful author, sportsman or musician.

Nozick also argues that each of us has a fundamental drive to be a certain sort of person. Let's say that in the experience machine I am clever, funny and charming. But even if everyone laughs at my jokes and nods admiringly at my sage observations, some part of me is aware that all of my experiences come from sitting in some sort of full-immersion tank, where my intelligence, humour and charm have no actual basis in my interactions with other real people.

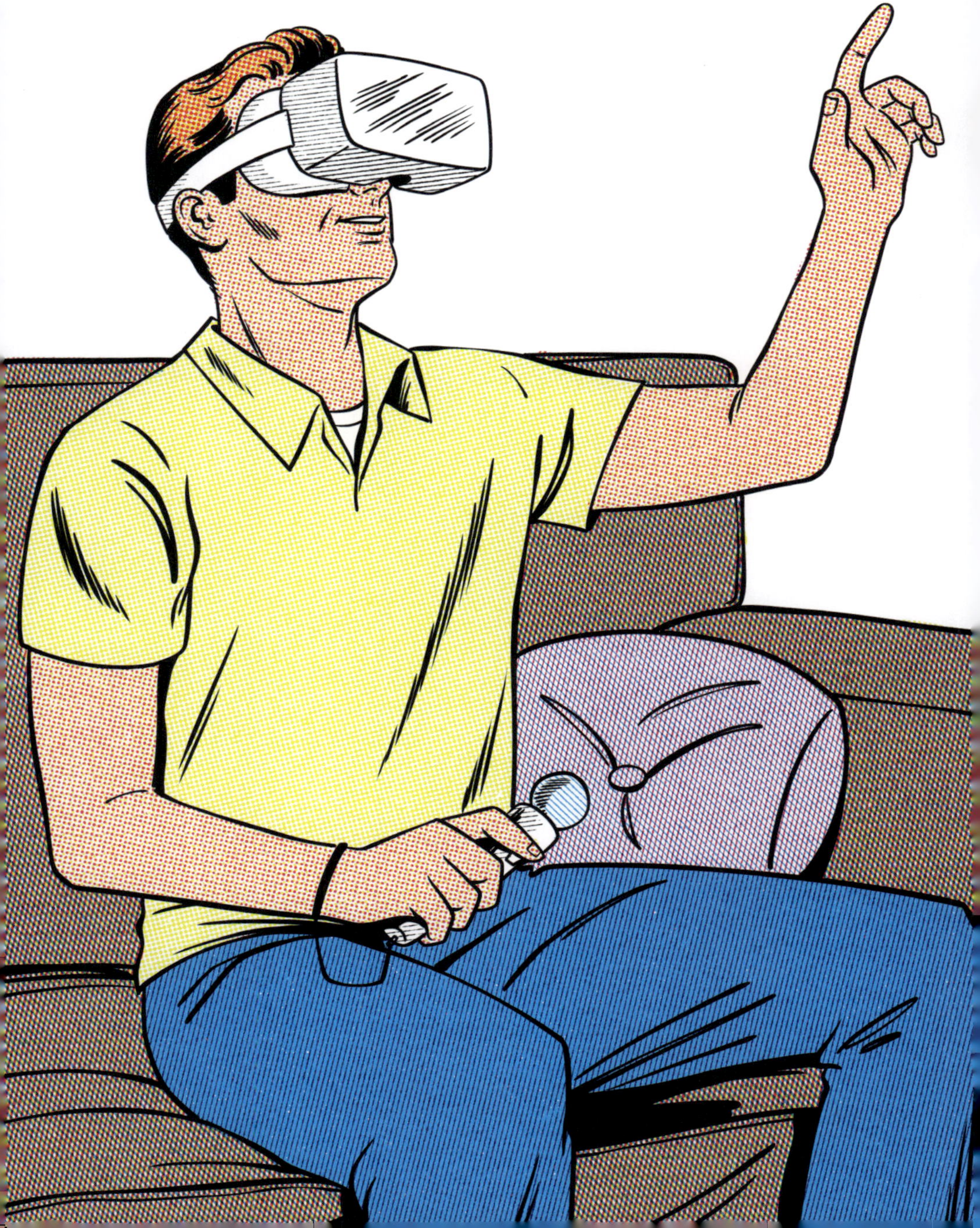

We can possibly think of other reasons to reject the experience machine, but most of them come down to the fact that what we experience there is simply not real. This means that, for most of us, arguably, truth is more important than pleasure. At least, in the long run, after we've become jaded by the endless procession of blissful experiences. Of course, we could programme in some frustrations, try to make the virtual seem more real, but then it would no longer be a machine purely for pleasure. And this is the purpose of Nozick's argument: to illustrate that pleasure cannot be the chief good.

The position Nozick rejects is known as ethical hedonism, of which Bentham's form of utilitarianism is a prime example. As we've seen, Kant also rejected this approach, arguing that the only motivation for being good is a sense of duty to the moral law, so pleasure cannot be the defining feature of an ethical act. Similarly, Aristotle argued that we all aim to live the good life – what he termed *eudaimonia*. This word has been translated in various ways – happiness, welfare, flourishing – but he rejected the idea that it consisted purely of pleasure. This was because there were other qualities important to the good life that were not guaranteed to be free of pain.

For instance, a good person must have the courage to overcome fear in the face of danger or to protect those they love; a temperate and rational person must control their desires in order to respect the needs of others, and so on. Aristotle also argued that pleasure could involve base and despicable things, which might lead us to wallow in greed and lust like animals if left unchecked.

Moral goodness, then, must be distinct from pleasure. It might involve pleasure – we all feel good about ourselves when helping someone, and even making a hard decision can feel better than the alternative if we're convinced that what we're doing is right. But goodness cannot be reduced to pleasure. This was a point made by English philosopher G E Moore (1873–1958), who argued that we can always ask if a certain pleasurable action is right, which we could not do if pleasure is simply the chief good. We can't say: 'This is the right thing do to. But is it right?' But we can say: 'This is the most pleasurable course of action, but I'm not sure it's the moral one.'

That said, ethical hedonists are not as unsubtle as the above picture perhaps suggests. For instance, the Ancient Greek philosopher Epicurus (341–271 BC) argued that the cultivation of mild pleasures was the basis of happiness. He had in mind such things as conversation, the appreciation of nature and the consolations of friendship. This is ironic,

EPICUREANISM

given that his name gives us the word epicure to describe someone who takes particular pleasure in sumptuous meals. But Epicurus was primarily concerned with moderation and in cultivating only those pleasures that would foster mental peace and equilibrium, or what he termed *ataraxia*. A true epicurean would therefore be more likely to nibble on fruit and nuts, perhaps with a single glass of wine, than to order a four-course meal.

We can also argue that it is impossible to completely divorce pleasure from moral goodness, and nor should we try to. Pleasure is a sign that things are well – health, a degree of comfort, sufficient money to allow for the enjoyment of pursuits and hobbies, the love of one's friends and family. All of these can be part of a moral life and we need not be embarrassed or ashamed of them. As with many things, it is perhaps a question of balance. As Aristotle himself put it, the right course of action is often a middle way between extremes. In rejecting its chief role, Aristotle nonetheless did not deny pleasure a key role, for a life without pleasure is no life at all.

# CAN A DEATH ROW INMATE BE HAPPY?

**Let us suppose that you are on death row. It's okay, I'm not going to ask how you got there. Maybe you were stitched up or perhaps you had a really bad lawyer. Whatever the case, you are probably facing death in the imminent future. Appeals are still ongoing but your odds aren't great. How do you face your fate? Can you accept it with equanimity? Make plans and enjoy what time you have left? Can you be happy?**

Most people, I suspect, would say no. How can I be happy with the threat of death hanging over my head? Well, it's all really a matter of perspective. For – sorry to be morbid – we're all constantly living in the shadow of mortality. As we go about our daily routines, walking the dog, doing our taxes, eating a meal, we're all just an inattentive driver, a sleeping aneurism or a lodged fish bone away from meeting our maker. Death is not something that it is completely in our power to control or avoid, and it is – theoretically – never far away.

This attitude is a key tenet of Stoicism, a philosophy originating in Ancient Greece that migrated to Rome at the height of the empire. Thinkers such as Zeno of Citium (334–262 BC), Seneca (*c.*4 BC–AD 65), Epictetus (*c.*50–*c.*120) and the philosopher-emperor Marcus Aurelius (120–180) tried to persuade us that consciousness of death and our powerlessness in life in general, far from being depressing facts, can actually liberate us. In this sense, your death is no different to all those other things that cause you anxiety, from your feuding siblings to your wayward offspring and your diminishing bank balance. By admitting this and accepting that most things in life are out of our control, we can cultivate a sense of inner calm and emotional equilibrium, or what was termed *ataraxia*.

We've come across this term already in relation to Epicurean philosophy (see page 105), but unlike Epicurus, the Stoics believed this inner state was related not to the cultivation of mild pleasures, but rather to a rational detachment from

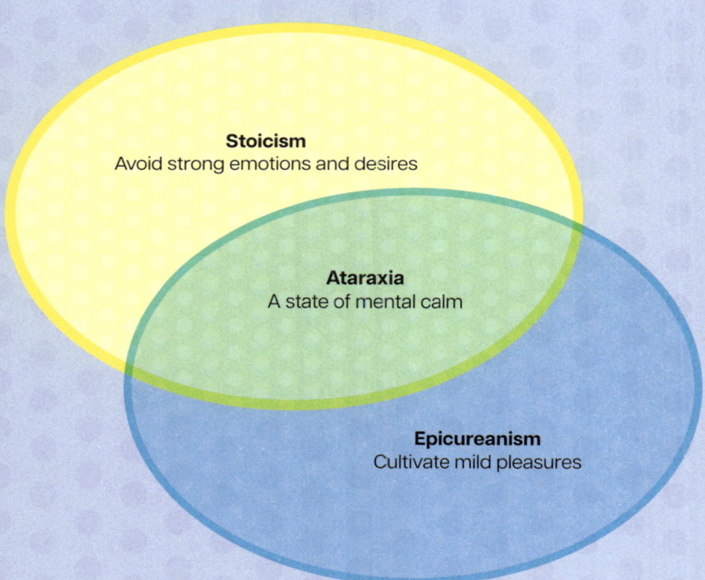


STOICISM VS EPICUREANISM

the things that might potentially cause us anxiety. What does it matter that my sisters are fighting, my child is off the rails or that my finances are down the toilet? Sure, I can try to amend these things – and the Stoics aren't saying that I shouldn't – but at the end of the day, they are all out of my control.

What *is* in my power is my mind and how I respond to challenges and turmoil. Key here is our attitude to fear, pain, desire and other things that can perturb us. By practising detachment, by realizing that the things I worry about are not essential to my happiness, I can cut the influence that these irrational powers have over me. Why be angry because my car won't start and I'm going to be late for a job interview? It's not like I'm a car mechanic and, even if I were, sometimes things just break, or you forget your toolkit.

But can we be happy under all circumstances? Even on death row? A lot of what we traditionally value in life was considered by the Stoics to be inessential to true happiness. While Aristotle thought that happiness required a certain minimum of the blessings of fortune – money, family, a degree of physical health – the Stoics denied this. They called these things indifferents, meaning they are things that are ultimately indifferent to our happiness. Yes, it would be nice to have them rather than not, for which reason some were termed preferred indifferents, but they are not essential. We can, if we must, be happy without them.

This seems like a hard ask – and it is – but what other choice do we have? Returning to Epicurus, he sought to console us in a slightly different way, arguing that the fear of death is based upon a misconception. We should not fear what we will not experience, and when we die, there is no conscious self with which to experience whatever waits for us in the post-mortem state. Fear of death is therefore irrational.

This is an interesting argument, but is it a convincing one? Setting aside the question of whether or not we experience anything after death, there is also the issue of what life we leave behind us. Perhaps our final days are made miserable by the thought of the feuding family we will leave behind, the loved ones we will miss or maybe just the fear of the pain that a protracted death might involve. Can we simply ignore all that? Again, it seems hard.

Bringing the debate up to more modern times, existentialism argues that death is a vivifying force, giving urgency and meaning to our actions. Thinkers such as the French philosopher Jean-Paul Sartre (1905–80) argued that there is no God, no fate or human nature that we can blame for our actions. If we did, it would be an act of what he termed bad faith. We have no choice but to be free and to face the anxiety of that responsibility head on. As scary as it may seem, this is a good thing and death is just a reminder that our dilemmas are finite, so we should get a move on and stop messing around.

THE MYTH OF SISYPHUS

Fellow French existentialist Albert Camus (1913–60) took a slightly different approach, arguing that the lack of any inherent meaning in life, coupled with our apparent need to find one, made our existence absurd. For this reason, his brand of existentialism is sometimes known as absurdism. He compared our daily struggle to find meaning with the eternal punishment of Sisyphus in Greek mythology. He was condemned by the gods to forever push a boulder up a hill, only to see it roll back down at the end of each day. For Camus, it is in our daily struggle against meaninglessness and the consciousness of our own mortality that we must find happiness. We must imagine that Sisyphus is happy, Camus argued, because it is in the full awareness of our predicament and the courage we find to face it, that any happiness must consist.

Again, not a very cheerful conclusion. But perhaps a necessary one.

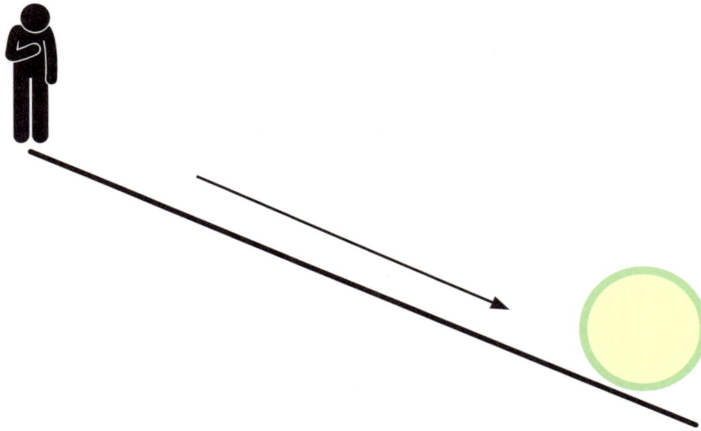

THE MYTH OF SISYPHUS

# CAN BAD PEOPLE DO GOOD?

**A man runs into a burning building to save the lives of a mother and two young children. His heroism hits the headlines and he becomes the toast of social media. However, all of this attention brings scrutiny to his private life and it emerges that he's the holder of some truly awful extremist views. Thanks to the internet, the news cycle is now much shorter than it used to be and today's hero can be tomorrow's villain. In fact, this very phenomenon has its own meme, called the Milkshake Duck. Aw! A cute story about a duck who drinks milkshakes! Oh, no. Turns out Milkshake Duck is a racist.**

But are someone's personal qualities relevant to their moral conduct? For different reasons, both Kant and utilitarians would say no. For Kant, as long as the man acted out of a genuine desire to do good for its own sake, his heroic actions should rightly be considered moral. For utilitarianism, it's only the outcome that matters, and since the man's actions increased the overall happiness of all those involved (and warmed the hearts of those who read the story), his objectionable political views are unimportant.

You may feel that there's something problematic about the idea of a bad person who takes credit for a moral action – and Aristotle would have agreed with you. In the *Nichomachean Ethics*, he argued that being moral is more about possessing virtuous characteristics than following rules, a position known as virtue ethics.

Plato, Aristotle's mentor, thought so too. He argued that an ideal individual should possess four key, or cardinal, virtues: wisdom, courage, moderation and justice, where the last is a sort of super-virtue that emerges from the possession of the first three. Aristotle built upon this, increasing the desirable virtues to a total of twelve (more or less). His list (see page 115) is perhaps not what a modern person would expect and included such things as modesty, wittiness and – perhaps the most dated – magnificence, which is the duty of a wealthy person to sponsor public buildings and temples. I guess we could rephrase this in terms of charitable giving on a big scale.

Of the three main ethical theories, only virtue ethics considers such virtues to be essential to moral action. True, Kant thought that there were good reasons to develop character and root out personal flaws, but that we did not have a strict or perfect duty to do so, unlike our duty not to

> Our characters are the result of our conduct.
>
> Aristotle, Nichomachean Ethics.

steal or lie. So, with imperfect duties, it's enough that we make some effort and we may choose which ones are most important to us. A utilitarian might also argue that, since it is unlikely that our attitudes and qualities have no impact upon others, then the possession of vicious characteristics may indeed have negative consequences. But it is only virtue ethics that makes the cultivation of good personal characteristics central to morality. The reason for this lies in the way it seeks to guide moral action.

Aristotle argued that one key, overarching virtue is practical wisdom. This is different from the virtue of wisdom conceived of by Plato, which was perhaps closer to intellectual knowledge (*sophia*). Aristotle valued that too, but he argued that practical wisdom is of vital importance to ethics, as it allows us to know when and how to apply this intellectual knowledge. Courage is good, but when should we run into a burning building and when should we wait for the fire brigade? Truthfulness is a virtue, but might there be times when it's prudent to stay silent? It is practical wisdom – what Aristotle termed *phronesis* – that guides us in the right action.

However, we are not born knowing this – unlike Plato, Aristotle did not think moral knowledge was innate – and so we must learn through doing. This means that childhood education is extremely important, as it forms the basis of our moral character and our teachers provide us with moral exemplars that we can follow while our reason is still developing, as we learn how most effectively to use this distinctly human quality to live the best life we can.

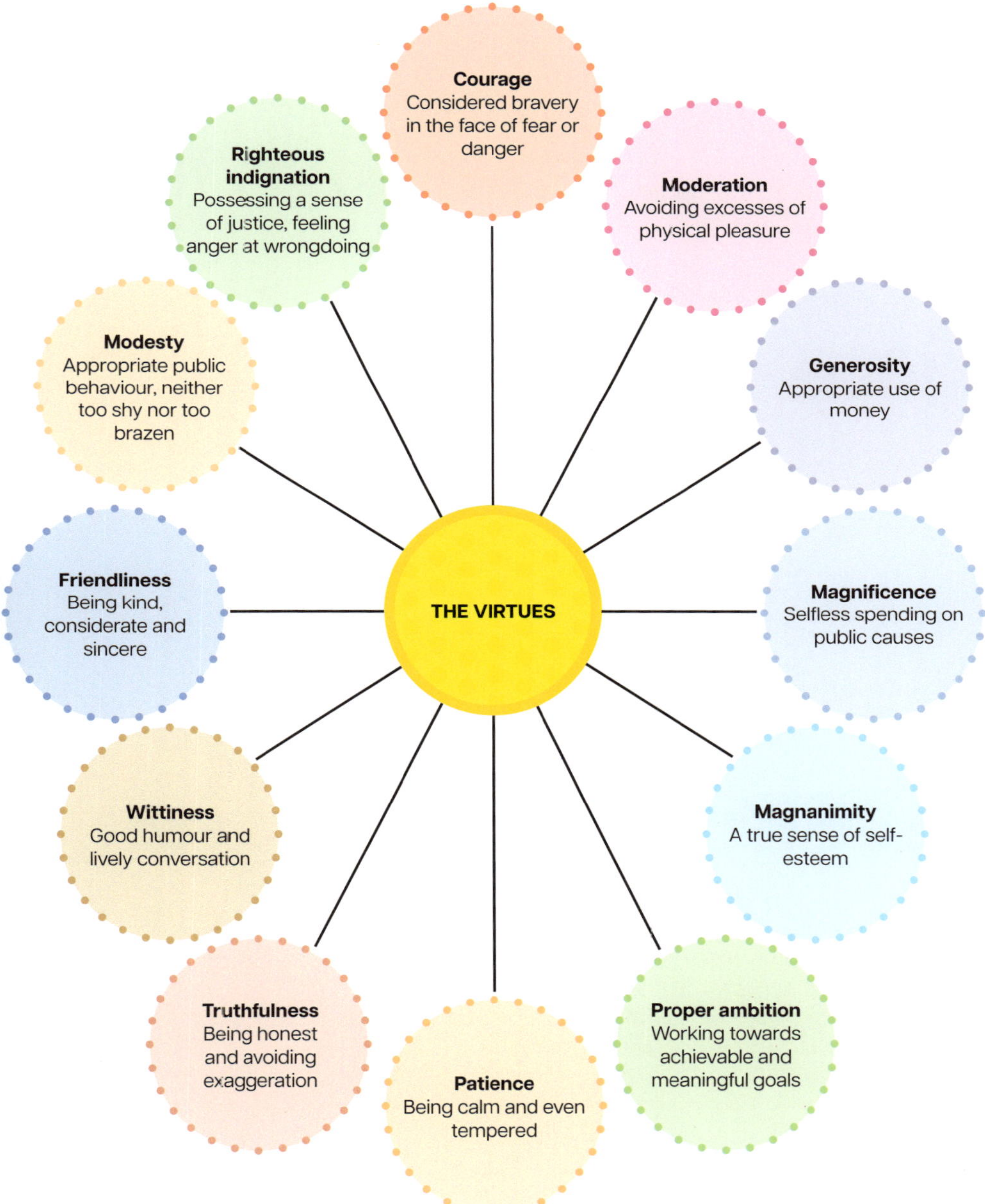


ARISTOTLE'S VIRTUES

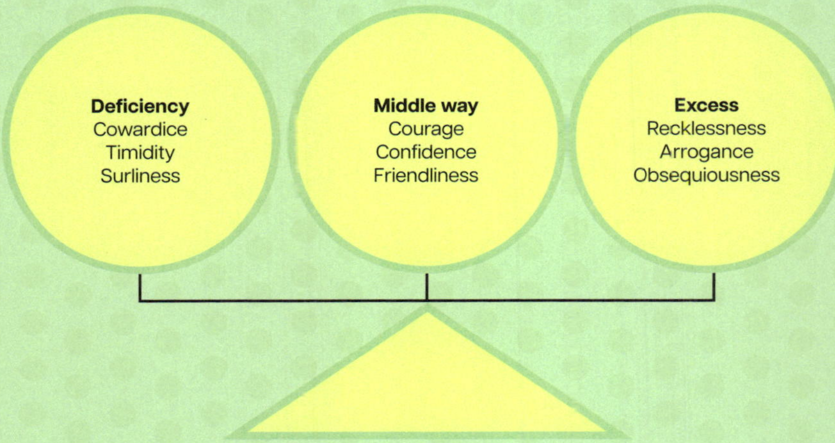


THE MIDDLE WAY

As noted earlier, Aristotle also thought that we may be guided by the general principle that the correct moral action often sits midway between extremes (see page 106). Generosity doesn't involve bankrupting yourself or being too stingy. Friendliness doesn't mean that you should be on good terms with everyone, including those of bad character, but nor should you be too withdrawn. This doctrine of the mean, as it's known, is therefore important. Unlike deontology or utilitarianism, virtue ethics does not have a strict set of rules that determine how we should act in certain situations. Its answer to most moral quandaries is basically to say that a generous/courageous/truthful person would act appropriately in this situation. Frankly, this isn't a great deal of help, so the doctrine of the mean serves as a sort of rule of thumb to guide us. Ultimately, for a virtue ethicist, what is moral comes down to the moral agent himself, his possession of the appropriate virtues and his application of practical wisdom. This is both its strength and its weakness, as what the theory gives us in flexibility, it loses in terms of specifics.

So what does this mean for Milkshake Duck and the have-a-go firefighter? Aristotle would say that, without possession of all the virtues, the man may not be properly motivated to act morally. Maybe his problematic views suggest bigotries and biases, and therefore a lack of justice in treating others fairly. So whether the act itself is commendable or not, the man as a totality is morally flawed. This seems harsh, and a high standard, but what it means is that we must judge someone as a whole, not from isolated acts of heroism or kindness. As Aristotle himself put it, a single swallow is not a sure sign that summer has arrived, no more than a single duck is a sign of...what *are* ducks a sign of?

# DOES RETAIL THERAPY WORK?

**Even the most cheerful of us get down from time to time and for some reason there's nothing like spending money (assuming you have money) to put a dimple in your cheeks (assuming you have dimples). I'm not sure of the exact reason for this. Maybe the thrill of acquiring something new and shiny, perhaps a nice shirt or a fancy phone, makes you feel on the outside how you'd like to feel on the inside. As I say, I'm not sure of the reasons – I'll leave that to the psychologists.**

Whatever the case, is there any harm in it? Lots of people have thought so. Isn't money said to be the root of all evil, after all? Well, technically, the love of money is the real problem (1 Timothy 6:10), so I suppose if you don't love money too much, or in the wrong way, then you're fine. But can you? Exercising such restraint has traditionally been seen as an act of moderation or temperance, and many of the ancient philosophers were big fans of it.

Plato, Aristotle, Epicurus and the Stoics all preached that happiness depends in part on our ability to control our desires. Plato personified these as a many-headed beast, its multiple mouths constantly clamouring to be fed. And he has a point. Many pleasures – those involving alcohol, gambling, drugs, sex – are insatiable and, if overfed, can lead to addiction, which in turn leads to misery. Can we add shopping to this list? Is shopaholism a thing? Whether it is or not, the pleasure you get from buying nice shirts or fancy phones is unlikely to fulfil you for very long, and a psychologist might argue that you are using it to mask some deeper issue that would be better faced directly. How *was* your relationship with your mother…?

This is not to say that no philosophers have been fans of money. The fathers of classical liberalism, including John Stuart Mill and John Locke, advocated not just for personal liberty, but for financial liberty and a move away from state control, and the two are often linked. Later, in *The Open Society and its Enemies* (1945), Karl Popper argued that

individuals living under communism, where everyone was considered equal and their needs were provided for by the state, were less happy than those living under less equal and caring democracies. This was because their lack of personal freedom made them miserable.

Friedrich Hayek (1899–1992) went further. In *The Road to Serfdom* (1944), he argued that an unregulated free market was essential to intellectual freedom and scientific progress, and that the commercial activity that it fostered was key in enabling people to live full and flourishing lives. These views have had a considerable influence over the last forty years or so, during which time the global economy has lurched from one financial crisis to another. Perhaps we now know enough to see that the case for such complete lack of economic regulation isn't clear cut. Without at least some safeguards, there is a real danger that laissez-faire economics (as this policy is called) can cause tremendous societal harm.

'But I just wanted to buy a new shirt!' you say. 'That's not going to cause a global financial meltdown!' Taken in isolation, maybe not. But can your freedom to buy whatever you want be divorced from the commercial system that gives you the opportunity? Let's say the shirt was manufactured in some poorer, developing country, because it's much cheaper for the brand to use the workforce there. And then let's say that not only are these workers paid less, but they also work longer hours, under poor health and safety conditions. And then let's say that some of these workers are children. Still getting that feel-good retail-therapy thrill?

The problem with freedom is that it is an emotive word. Who wants less freedom? Put like that, no one. But we must always ask what that means in practical terms. 'No more red tape!' sounds like a great slogan, but what if that red tape involves forbidding children from working long hours in dangerous conditions for little pay? Is that the sort of red tape you want to be free from?

You may say it's hard to shop ethically. True – especially when many manufacturers are using the same exploitative supply chains to keep costs to a minimum. However, consumer activism can be a thing. If there's enough public momentum to boycott a product or company, then it really can effect change. Most companies now only have ethical and green policies because enough individuals refused to buy products made under exploitative conditions and some decided to make a fuss about it. There's nothing like bad press to make the shareholders jittery.

Such developments also mean that you are now more likely to find ethical alternatives to the big companies, as smaller suppliers step in to address the growing demand for responsibly sourced goods. The internet is great for that. Yes, you may end up paying a bit more for that shirt or phone, but you can at least be sure the extra money is going to support ethical policies that don't make you feel ashamed. In fact, they might give you a good feeling inside, like you're making an actual difference. Which is a form of retail therapy, too, I suppose.

# ART, BEAUTY & CULTURE

## AESTHETICS

# 4.

# WHEN IS A TOILET NOT A TOILET?

**Answer: When it's a work of art. Perhaps surprisingly, this is not the punchline to a joke. Artists have long been pushing the boundary as to what is traditionally considered worthy of artistic merit.**

British artist Damien Hirst (1965–) famously pickled a sheep in formaldehyde (*Away from the Flock*, 1994). Before that, in 1961, Italian artist Pietro Manzoni (1933–63) assembled 90 tin cans labelled *Merda d'Artista*, allegedly containing samples of his own poop. So, what's stopping you from uprooting your toilet and schlepping it to the nearest gallery? Well, apart from the fact that it might leave you inconvenienced, someone has already beaten you to it.

In 1917, the French painter and sculptor Marcel Duchamp (1887–1968) created a piece he called *Fountain*, which consisted solely of a conventionally manufactured porcelain urinal that he had signed and dated 'R. Mutt 1917'. The piece caused a controversy and was refused a place in a gallery in Switzerland where it was submitted for display, but a year later became a centrepiece at an exhibition in New York.

Duchamp went on to create other 'readymades', a term he used to designate found, everyday objects that he would customize in some small way, often to give them a humorous or anarchic slant. As a result, Duchamp became associated with *Dadaism*, an anti-art art movement (if there can be such a thing) that focused on pricking the pretentiousness and elitism of the art establishment. It is not a little ironic that Duchamp has since gone on to be considered a founding figure in the world of avant-garde conceptual art and has inspired many other artists (Hirst included) to follow in his footsteps.

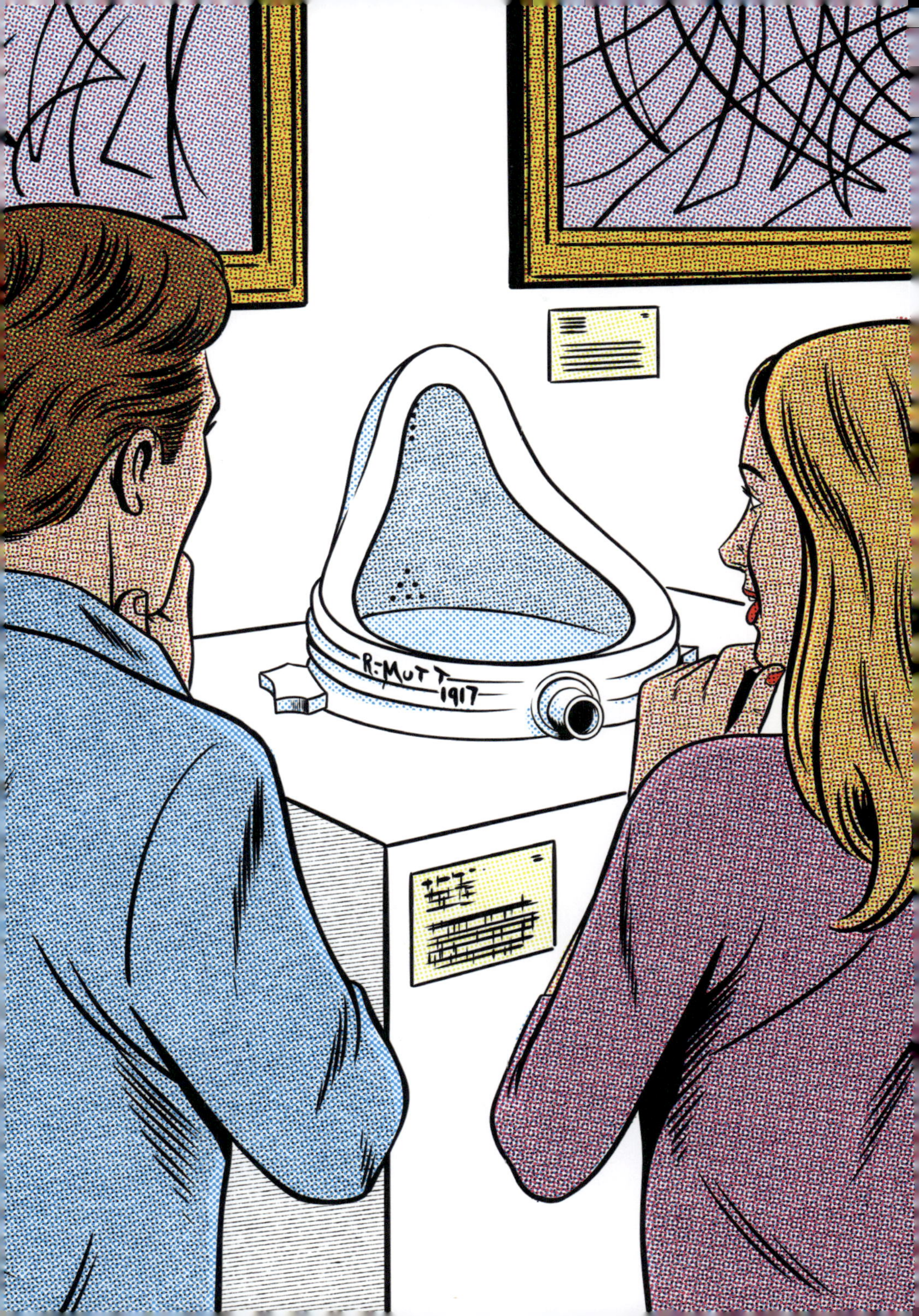
R. MUTT
1917

This shift from outrage to acceptance is not new. The Impressionist works of Claude Monet (1840–1926), Edgar Degas (1834–1917) and Auguste Renoir (1841–1919) are now considered central to what people commonly think of as great art, but at the time the style was rejected by the art establishment – 'Impressionism' was originally coined as a dismissive, critical term. I won't say that everyone now accepts toilets as works of art, but today's art world is more receptive to pieces that are intended to mock or shock it, to the point where it's perhaps increasingly hard to do so.

Take, for example, *Girl with a Balloon*, a framed piece by the street artist Banksy, which, immediately upon its sale at auction in 2018, started to shred itself via a secret in-built mechanism. This only half-worked and what was perhaps the artist's attempt to satirize the greed and shallowness of the contemporary art world backfired – when half-destroyed, the piece became even more valuable.

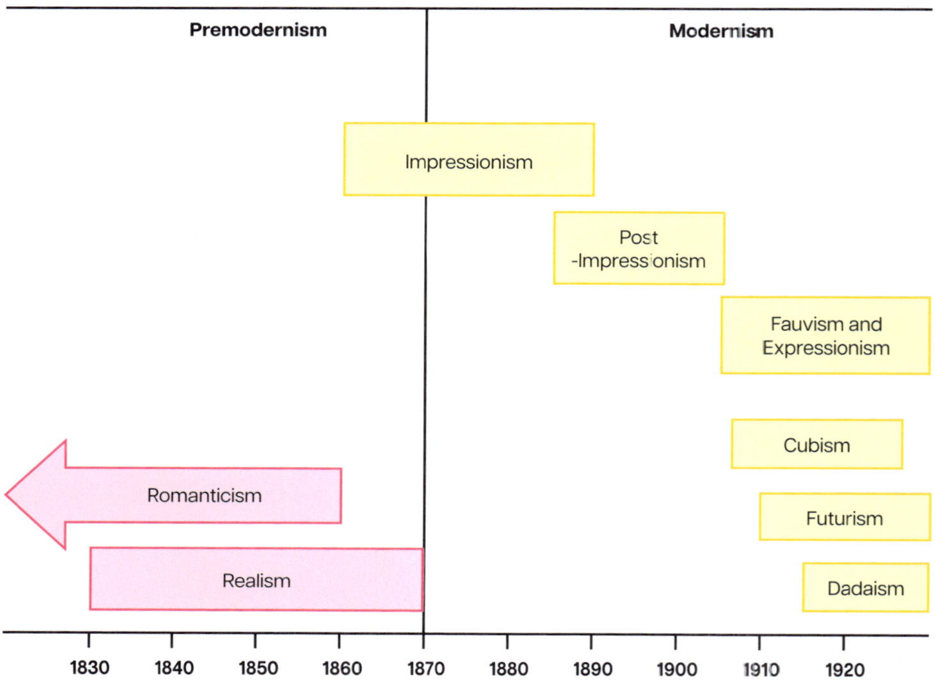

ART MOVEMENTS

Over time, art has moved away from embodying traditional values and skills towards a more conceptual approach. Plato argued that great art copies or represents the truth, which suggests that great artists are those that are most skilled in creating that resemblance. From the Renaissance onwards, the lustrous colours made possible by new pigments and oil paints encouraged artists to appeal to the senses, depicting human and natural forms in exquisite and sumptuous detail. In reaction to the invention of the camera, Western art began to move away from realism and became a vehicle for subjective emotion and vision through the works of such artists as Vincent Van Gogh (1853–90) and Edvard Munch (1863–1944).

Later still, the English art critic Clive Bell (1881–1964) argued that what was important in art was its form, or the way in which the various elements composing the artwork were arranged. This in turn led to abstraction, a further

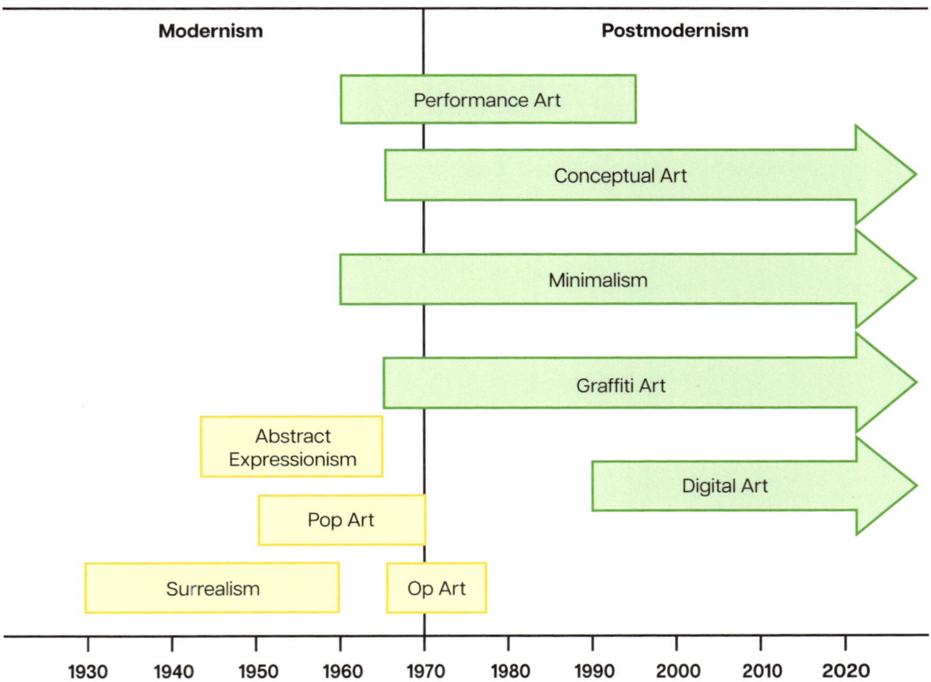


ART MOVEMENTS

move away from realism, as artists such as Piet Mondrian (1872–1944), Wassily Kandinsky (1866–1944) and Mark Rothko (1903–70) attempted to capture moods and ideas without the use of representational figures.

While form, composition, emotional vision and representational skill are still important considerations for many contemporary artists, they no longer define what art is. So, what does?

The institutional theory of art was independently proposed by philosophers Arthur Danto (1924–2013) and George Dickie (1926–2020) during the 1960s and 1970s. They argued that art is something that has been created and has had the status of 'artwork' conferred upon it by suitably qualified people – that is, the curators, dealers, gallery owners, collectors, writers, artists and critics that together make up the art world.

This seems a sensible theory – the best-placed people to inform us about art are surely the people who appreciate and practise it – but it does come with a downside. Chief among these is perhaps that it makes things a bit insular and arbitrary: art is simply what the art world says it is. As such, an artist's work might display any number of features that would traditionally have qualified it as art (skill, composition and vision, for example), but still be excluded or dismissed. This invites the question: is the art world an esteemed group of experts, or simply a cabal with its own whimsical set of values?

The problem is compounded by the philosophical and cultural movement known as postmodernism, which rejected the traditional values and theories that had previously informed art criticism. Philosophers such as Jacques Derrida (1930–2004) and Jean-François Lyotard (1924–98) have argued that there is no such thing as objective truth, no independent ethical or aesthetic principles, and no standards of justice to which we can all appeal. Lyotard rejected such metanarratives, or overarching viewpoints, arguing that all we have are local narratives, or culturally biased, arbitrary traditions.

Similarly, Derrida claimed that there exists only the assortment of competing positions that make up the world, each of which may only be defined by their *différance*, or relative variation from one another, and not in relation to some independent standard. In other words, no one really knows anything, because our views are all unconsciously shaped by our own inherent prejudices and blind spots. Even science is just another way of seeing the world.

When applied to art – where postmodernism has become hugely influential – you can perhaps start to see why, from this point of view, signed toilets and cans of artist poop are no less art than a painting by Degas or Monet. You might not agree with that, but...*who* did you say you were again?

# WHO WROTE THIS BOOK?

**Imagine a famous writer – a dead one. He is celebrated throughout the world. Tourists come from foreign countries on pilgrimage to pay homage to his place of birth, and his face appears on T-shirts, tea towels and novelty mugs. His works are compulsory reading, taught in schools and studied in universities, a cornerstone of culture, and almost everyone knows a phrase or two of his they can quote by heart. Anyway, you get the picture. He is Famous with a capital F.**

Now, imagine that you are an antiques dealer and one day, many years after the famous author's death, you are digging through a dusty old house in his home town. You're looking for undervalued treasures, hoping you can pay the nice old lady who owns the place less than they're worth, when you come across a box of old books. Actually, they're really old – the pages are browned and brittle, mottled with age – and among them is a diary. The famous writer's diary. It's a confession, revealing that he was not the sole author of the books for which he is celebrated, rather that they were written by a committee of talented writers who wished to remain unknown. The group paid him to act as a frontman and to keep their identities secret. As you read on in astonishment, you begin to realize that the true message of the famous man's books is far different from what everyone previously thought.

The above scenario is a sort of metaphor for how we look at authorship. When we consider a book written by a single individual, we generally assume that he or she had a specific meaning in mind – one which we, as the reader, can uncover, as if we're detectives. But is this the case?

Russian literary theorist Mikhail Bakhtin (1895–1975) explored this question and argued that any language is what he called a heteroglossia, containing within it numerous other languages or points of view, which compete with and overlap each other. To try to impose a single viewpoint – such as in a system of linguistics – is therefore arbitrarily to favour one voice over another. When we apply this to the novel, Bakhtin argues that it is polyphonic, because the author will often employ different narrative voices, styles and kinds of language to express different themes and characters. In doing so, the author's own voice becomes intermingled with that of others, until we reach a point where we might consider the author as no more than a mouthpiece for this dialogue between different points of view – a sort of ventriloquist, perhaps.

Even if there was a single authorial point of view behind a text, could we know it? This is a point made by W K Wimsatt (1907–75) and Monroe C Beardsley (1915–85), in their 1946 article, 'The Intentional Fallacy'. They argued that it is a misconception to think that we can divine the true intentions of an author based on an examination of the text. How do we know what any person truly intends or thinks by what they say?

And even if we could know that, should we judge a work of art on the intentions behind it, or rather whether it works as a poem, novel or painting? The latter seems more important and we should therefore evaluate a work of art using internal evidence, such as how effective its techniques are and whether it is skilful, original and insightful. From such a point of view, the antiques dealer's discovery of the diary is irrelevant.

In his 1967 essay 'The Death of the Author', French philosopher Roland Barthes (1915–80) went further than this, arguing that it is the response of the reader that is most important and the author's intentions are immaterial. The moment a text is published, it belongs to the reading public and its meanings become plural. So, to try to give a single, 'authorized' interpretation is to limit its meaning and interpretative richness. Each text has multiple possible meanings and it is therefore a form of tyranny to impose just one. This means that to all intents and purposes, the author doesn't exist – is 'dead' – and there are only readers.

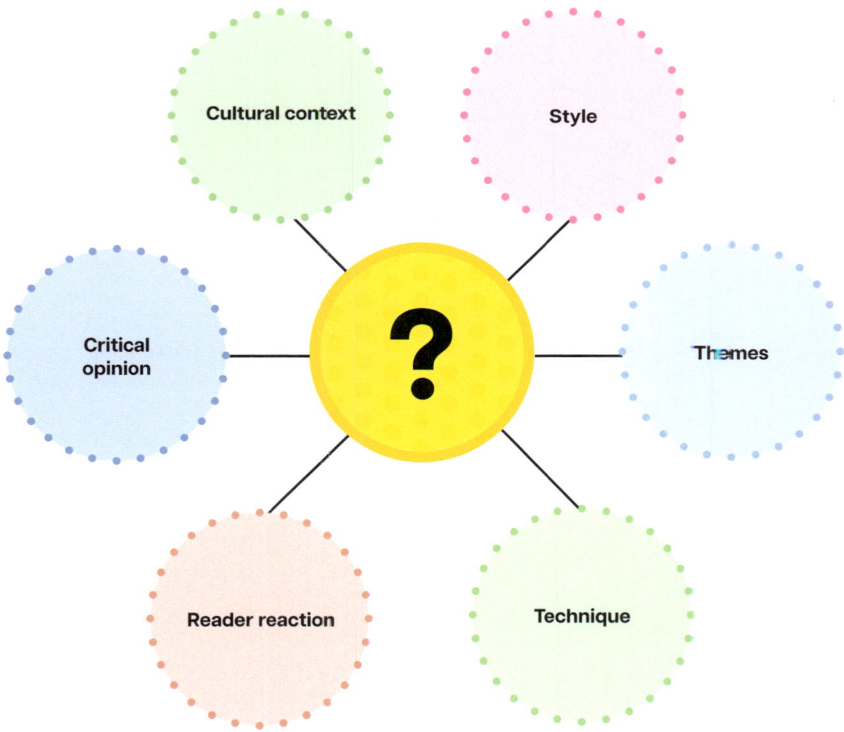


WHAT SHOULD WE CONSIDER INSTEAD OF AUTHORIAL INTENTION?

This approach perhaps reaches its most extreme expression in the work of French postmodern philosopher Jaques Derrida, who argued that 'there is no outside-text' (*Of Grammatology*, 1967). This has sometimes been interpreted as 'everything is language', but this is not true. He's actually saying that there is nothing that gives any external authority to an individual viewpoint. As I've said previously, for Derrida, God does not exist, truth is not some objective standard to which we can appeal and even science is not objective (see page 129). Rather, each viewpoint contains its own in-built justifications and we must deconstruct it to understand how these inherent biases and prejudices give the illusion of authority. For this reason, Derrida's approach is sometimes known as *Deconstruction*. Which would seem to suggest that I can't even claim to know with any certainty what I meant by my own book.

As you can see, the question of authorship is a complex one. Even if I am, unlike the above scenario with the famous writer, entitled to be considered the author of my own book, there is a sense in which the language I use doesn't fully belong to me. In the act of speaking or writing, I must use words, phrases, idioms and metaphors that have acquired their meaning communally. The author is therefore just a convention, a way of speaking. And if the author doesn't exist, then what are we to say about the self?

# WAS SHAKESPEARE ANY GOOD?

**You're at a party – the sort where people eat canapés, try to say clever things and impress others with the books they've read – and someone is holding forth about Shakespeare. Maybe it's a glass too many of the complimentary wine or maybe it's a decade's worth of having *Romeo and Juliet* rammed down your throat at school, but you find yourself coming clean: you don't like Shakespeare – never have. You think he's boring, melodramatic, wordy and frankly overrated. You'd rather watch *The Real Housewives of Beverly Hills* than *Hamlet* or *Henry V*.**

The room goes silent. In the distance, a dog barks. Someone titters nervously. Have you made a dreadful faux pas? Or just said what everyone else is too afraid to admit? You finish your wine and leave.

It might seem silly to question the status of a literary figure whose genius has been celebrated the world over for centuries, but you wouldn't be the first. The great Russian novelist Leo Tolstoy (1828–1910) thought him tedious, deficient in artistry and even lacking a sense of morality. Fellow playwright George Bernard Shaw (1856–1950) thought Shakespeare's works superficial and the man himself a vulgar snob.

Maybe you're not wrong in preferring reality TV to *Richard III*. After all, as the saying goes, there is no argument in matters of taste. Anyone can like whatever they want, can't they? Isn't all taste subjective? You wouldn't be criticized because you don't like cabbage but you do like carrots.

Well, this is true to an extent. David Hume argued that we may divide opinions into two: judgements and sentiments. A judgement concerns a matter of fact – whether it is going to rain later today, or whether it rained last Tuesday. Judgements can be wrong – you might forget something, make a mistake or rely on faulty evidence – but when you judge something, there is an accepted way of determining the truth by appealing to logic or evidence.

Sentiments, on the other hand, are feelings or emotions, so while these can be based on false judgements (your jealousy is baseless because your partner's not cheating after all), in terms of mere liking or disliking, we can't appeal to anything outside of them to see if we are right. You recommend banana ice cream to a friend and they hate it. Are they wrong? No, they simply have a different reaction from you. The same can apply to all manner of things – comedy, music, sport, clothes. Maybe we should all just live and let live, and leave it at that.

It's worth noting that taste is used broadly here to mean subjective opinion. Hume is not confining himself to talk of actual taste in food and drink, although he did see a connection between aesthetic sense and gustatory preferences. While he agreed that taste was subjective and often differed in each person, he also noted that it is common for people to agree about what is good or bad. This therefore suggests that there are in fact standards of taste that we can appeal to. But what are they?

Hume identified five main criteria, through which we may become, in his phrase, 'true judges' or skilful critics.

**1** The first is that a sound appreciation of an artwork must be based on good sense, that is, the use of reason in identifying certain principles and criteria and then applying them. So, for instance, we might look to see whether a painting is well composed or displays a skilled use of colour.

**2** A good critic also has a refined sense of sentiment, or delicacy, and can distinguish between the expression of coarse or fine emotions, or between the heavy-handed or subtle use of themes.

**3** Taste is not something we are born with; we must train ourselves to acquire it. Therefore, another key element in developing our taste is practice, as we apply these skills through repeated exposure to a variety of works of art.

**4** Through practice we may compare different pieces and better distinguish between varying degrees of skill and artistic vision.

**5** Finally, we must learn to suspend our own prejudices – both positive and negative – and view novel and unusual works with an open mind, as well as setting aside the personal like or dislike we might feel for an artist we already know.

In summary, a skilled critic will therefore possess a 'strong sense, united to delicate sentiment, improved by practice, perfected by comparison, and cleared of all prejudice' ('Of the Standard of Taste', 1777). If we can do all of this,

Hume argues, then we will see that while all forms of taste are in a sense subjective, they are not merely arbitrary, for there are widely held standards that we can appeal to in judging something's worth. Does that mean you're wrong about Shakespeare? The fact that there are near-universal aesthetic standards does not mean that we cannot have differences of opinion about art and culture or food and drink – we certainly can. And, in fact, detractors of Shakespeare have generally appealed to commonly accepted standards in order to criticize him. But maybe the fact that so many agree that Shakespeare was a genius means that negative opinions of him are less respected than the positive opinions of the majority. Perhaps, as long as you can make a rational case by appeal to such standards, you are just in the minority. There's nothing wrong with that. And who knows, tastes might change.

This highlights some exceptions that Hume notes about standards of taste. Cultures may differ across time and geography, and we may find that accepted standards do too. We may also find that personal tastes vary. Despite all the reasoned arguments that his supporters present and the admitted skill and artistry that he undoubtedly displays, you still just don't get Shakespeare. He doesn't do anything for you. I understand. I feel much the same about ballet or opera – two extraordinarily skilful and impressive activities, which nonetheless leave me cold.

But hey, there's a new season of *The Real Housewives of Beverly Hills*, if you fancy it?

> 'One man's meat is another man's poison.'
>
> – Lucretius, On the Nature of Things

# IS ART JUST PROPAGANDA IN DISGUISE?

**The world is awash with politics – you can't escape it. It's not that you don't care, it's just that, with all the stresses of life, sometimes you simply need a break. So, whatever you're into – romance novels, perhaps, or science fiction – it's nice to occasionally just lose yourself in the fantasy world of a book or a film, to escape all the discourse and engage in imaginative fun. Except, if Karl Marx was right, no such escape is truly possible.**

The reason for this, according to Marxist theory, is that the values that determine the shape of society stem from the economic base or substructure. The form this takes depends on the mode of production. This includes the means of production (the resources, tools and technology used in producing things) as well as the relations of production (the social relationships that govern these activities).

If you live in a capitalist society, economic activity might involve farming, mining and other uses of the land, as well as the manufacture of goods, the provision of services and the teaching of skills and knowledge. Of course, many of these activities also take place in other types of society – people in tribes or communes also need to eat – so what shapes the base in a specific way is the relations of production. In capitalism, this usually takes the form of employer and employee, where the former (the bourgeoisie or capitalist class) owns the means of production (factories, offices and technological infrastructure, for example) and pays the latter (the proletariat or working class) a wage in return for their time and labour.

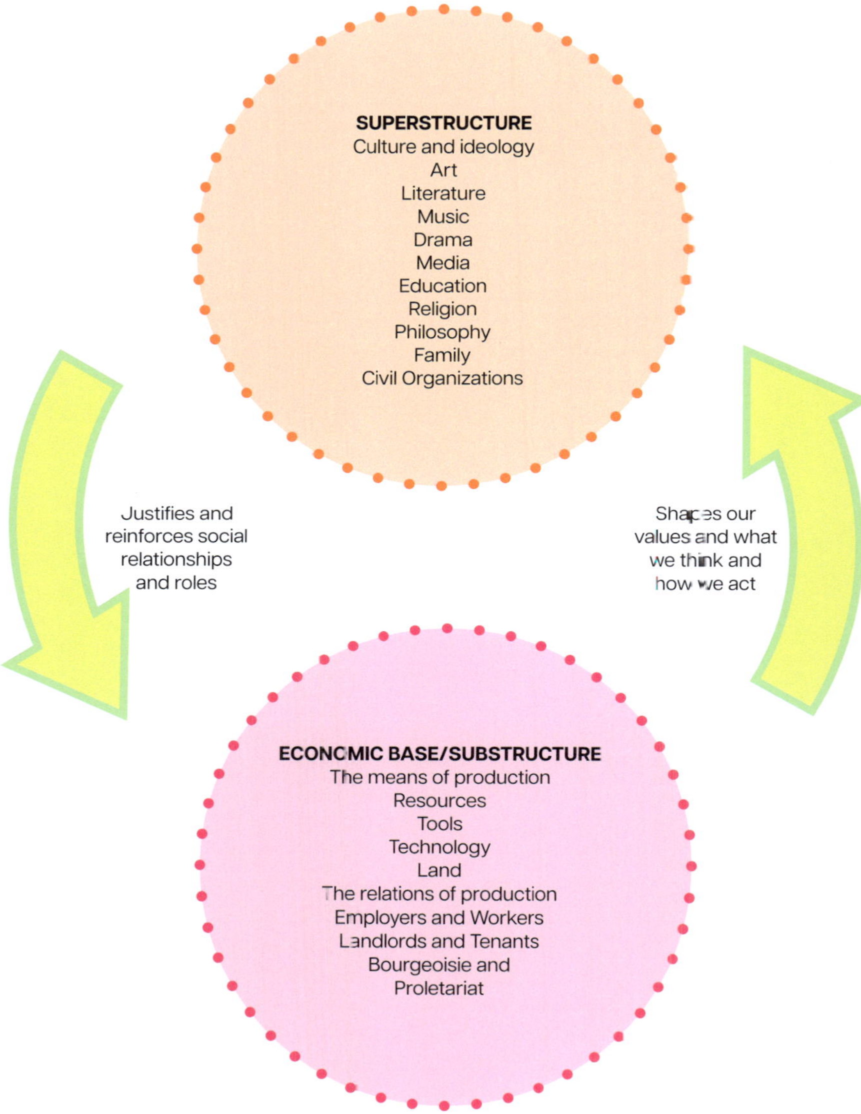


BASE VS SUPERSTRUCTURE

Of course, capitalism is not without its critics. Marx himself thought that it was exploitative and alienated the worker from the products of their labour. Once, a rural worker might have laboured for their own subsistence or for that of their community by hunting and gathering, maintaining a small crop or vegetable plot. Under capitalism, they are wage slaves, forced to earn money for life's essentials in return for producing something for which they may have no personal use or the money to afford.

In a capitalist society, there is little opportunity for the worker to own their own means of production (the farm, factory or office where they work), so the worker is often forced to accept the wages on offer, which are kept low by market competition. By contrast, as the owner of the means of production, the capitalist proprietor can reap profits on their investments by paying workers less than they are truly worth, squeezing out of them what Marx called surplus value.

Whether you agree with this analysis or not, you can see that such a society could not continue without some sort of justification. If the system is so unfair, then what is stopping the workers from rising up and overthrowing it? Either they agree with the values their lives embody or something else is keeping them in their place. For Marx, it was the latter, which is where the superstructure comes in.

Outside of work, a person's life is influenced and shaped by social institutions, including schools, church, the media, hospitals and family. Through these, the message of the ruling classes and business owners is reinforced and justified. This is why Marx termed religion 'the opium of the people' – it encouraged the masses to focus on immaterial matters and helped to lull them into accepting the way things were. They were led to believe that if they were unhappy, it was due to their own spiritual deficiencies or leading a life of sin. The very desire for a better material existence might be framed in terms of the sin of greed – and isn't poverty a virtue?

The Roman poet Juvenal (*c.*55–*c.*128 said that the masses would be happy as long as they had 'bread and circuses'. Marx also recognized the diversionary effect that cultural

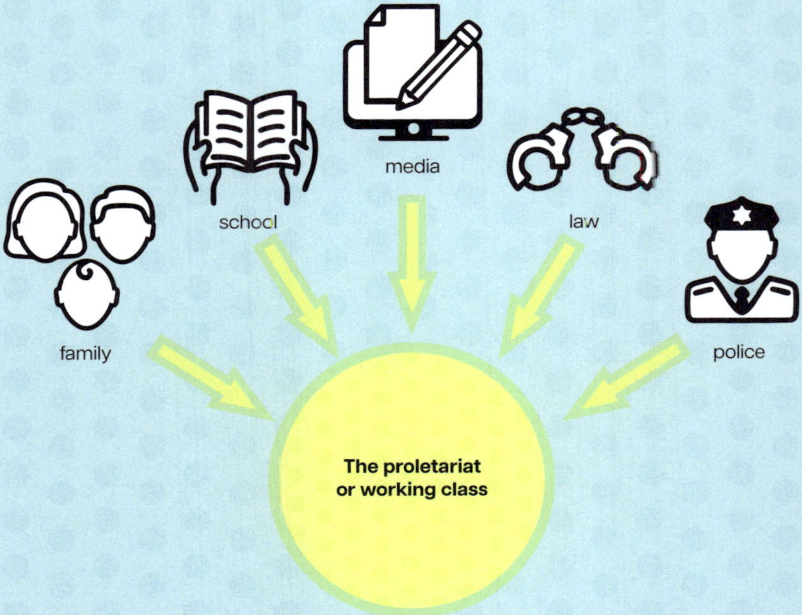


CULTURAL CONTROL

pursuits could provide. At the same time, he argued that, while the values of the dominant classes were drummed in through church, school and workplace, they were also reinforced through literature, theatre and art.

These values could be transmitted in different ways. Rudyard Kipling (1865–1936) deliberately championed the virtues of the British Empire, while even Jane Austen (1775–1817) effectively acted as a vehicle for cultural indoctrination. In all of her novels, Austen focused on social and romantic comedy. Only once (in *Persuasion*, 1817) did she mention the Napoleonic Wars (1803–15), the existence of which must otherwise be inferred from events, such as the dastardly Lieutenant George Wickham in *Pride and Prejudice* (1813) being stationed at a military camp in Brighton. Even this non-political focus reinforces the values of the society it describes. In *Emma* (1815), while the titular character idealistically tries to matchmake between people of different social classes, the end result is often conventional.

Emma herself ends up with the economically and socially appropriate Mr Knightley. As for slavery, which is the unspoken basis of the wealth of a few of the characters in the novels, it doesn't get a mention.

Okay, then what about fantasy – *that* can't be politically motivated, can it? Well, take J R R Tolkien's *Lord of the Rings* (1954). Many of the good characters are associated with traditional, class-based values. The peace-loving hobbits spend their lives of rural medievalism under the guidance of kindly wizards and lords, while the aristocratic elves have refined and elitist attitudes.

In contrast, the bad guys – the orcs and the followers of the Dark Lord Sauron and the evil wizard Saruman – are associated with working-class culture and industrialization. Whether consciously or not, Tolkien is making it clear where his allegiances lie. And unless you are careful, dear reader, as you root for the plucky Frodo and the noble Aragorn, you will find your sympathies unwittingly drawn into alignment with the author's own subtly concealed religious, political and cultural values.

But is it all so simple – must all art have a political message? What if a writer rejects the Marxist interpretation of their work and says that's not what they meant? In response, the Marxist can simply claim that while the author may not have been following a conscious agenda, they were nonetheless an unwitting mouthpiece for capitalist values. As another Karl pointed out – Karl Popper, this time – there's no way to prove the theory wrong and the Marxist is right either way. Which, Popper concluded, isn't exactly very scientific.

# IS A MOUNTAIN BEAUTIFUL?

You are walking in the countryside when you see a lovely flower standing out amid the grass at the side of the road. It is so perfectly formed and intricately coloured, and the sunlight hits it in such a way that it glows as if with an inner force. It is beautiful. You carry on walking with a smile on your face and a spring in your step, but as you turn the corner you are presented with a view that gives you quite a different feeling.

Enormous, cloud-capped mountains stretch up before you, towering over a deep valley that seems to plunge down into the very depths of the Earth. Is it beautiful? Undoubtedly. So why do you feel a bit sick?

The traditional notion of beauty is that it is a sensory experience that gives us pleasure. Plato discusses it in several of his dialogues, where it is central to his philosophy as a whole, being linked to both truth and goodness. It is our reason that discerns the form and structure in the sensory experience of beauty, which is why ugly things are formless. In his 'Ode on a Grecian Urn', the English Romantic poet John Keats (1795–1821) agreed, writing: 'Beauty is truth, truth beauty'.

However, in the 1st century AD, the Greek writer Longinus argued that there are certain thoughts and emotions that go beyond our ordinary experiences and which elevate us, causing us to bypass reason and adopt feelings of awe and reverence. This notion of the sublime in art and writing was picked up by later European writers from the 16th century onwards, when it became an important concept in literary theory and aesthetics. Perhaps the most famous and important expressions of this idea are to be found in the writings of the philosophers Edmund Burke (1729–97) and Immanuel Kant.

In *A Philosophical Enquiry into the Origin of Our Ideas of the Sublime and Beautiful* (1757), English philosopher Edmund Burke argued that the beautiful and the sublime were two distinct emotions. Burke agreed with Plato that the beautiful concerns things that give us pleasure, such as well-formed, delicate and harmonious objects like the flower you passed, which we are drawn to out of love and affection.

In contrast, the sublime is something that produces in us a feeling of fear, dread, reverence or awe, due for example to its incredible size, age, power, obscurity or terror, which makes us feel overwhelmed, even though what we experience may not actually be a source of direct danger to us. In other words, the sublime is an emotional tension between admiration and repulsion.

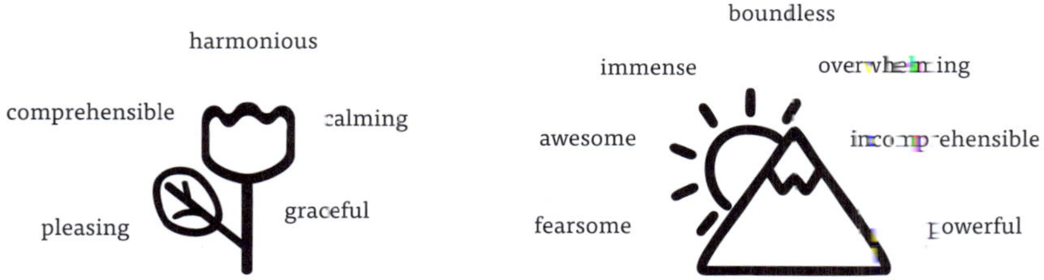


BEAUTY VS SUBLIME

Kant discusses the sublime in various works. In *The Critique of Judgement* (1790), he argues that what differentiates it from beauty is its form. Beauty has boundaries (a limited shape and size), whereas the sublime is boundless. It is this quality that overwhelms our understanding and imagination, either in terms of size or time (what he called the mathematical sublime), or in terms of its power (the dynamic sublime), leaving only our ability to process the situation through our intellect.

We can conceive of infinity or eternity in abstract terms – it's just something that goes on without limit – while still feeling emotionally disturbed and belittled. So, your view of the soaring mountains and plunging valleys takes your breath away while making you feel slightly afraid. This is even though neither the mountain nor the valley are of any immediate danger to you – assuming the mountain isn't a volcano on the point of eruption and there is no danger of you falling into the bottomless valley.

There is an important distinction here that may better help you understand what Kant is saying. The sublime experience is something that is partly incomprehensible to us. This is because it is the understanding, that part of our mind that attempts to comprehend the information processed via the senses, which is bamboozled. We see something or hear someone speak and we attempt to understand or make sense of it.

In contrast, the reason or intellect deals with abstract ideas. So, intellectually, we can process ideas (such as things of infinite size, age and power) that we may at the same time find incomprehensible when considered only via our understanding and imagination, because we are able to reflect and reason about them.

We can see here how useful this distinction is, as it helps us to understand not only our attraction to beautiful things, including flowers and faces, but also why we are drawn to things that terrify us, including horror films and jaw-dropping cosmic vistas. Our reason allows us to overcome our fearful and awe-struck reactions and achieve a more detached and elevated perspective, where we can experience the majestic thrill safely.

A similar point was made by the German theologian Rudolf Otto (1869–1937) in *The Idea of the Holy* (published in German in 1917 and translated into English in 1923). Otto argues that the term 'holiness' contains more than just a sense of moral perfection and also covers certain religious or mystical feelings that originate outside of us. In this sense, we may encounter the holy in a way that disquiets us or makes us feel awe or dread.

Otto's term for this is the numinous, which was influenced by Kant's concept of the noumenon, which refers to the aspects of reality that are beyond the limits of human reason and experience. It is therefore the feeling in us that there is some inconceivable power or force beyond reality as we know it that creates in us a sense of unease, awe, dread or even terror. It is not necessarily that such a divine or cosmic force bears us ill will; it is that its sublime qualities represent something wholly alien and other.

# AM I IN LOVE WITH MY BEST FRIEND?

**You've known each other for years – you grew up together. You like a lot of the same books and films, and when either of you has news or is upset, you call each other first. Then your friend gets a romantic partner – which is fine. It's great. You're happy for them both. But suddenly, the calls and the texts aren't so frequent, and the two of you don't hang out so often any more.**

That's understandable, as new relationships tend to do that. But while you understand, you must admit that you're finding it hard. You feel a little abandoned, truth be told, a bit betrayed. Even a bit…jealous? *Oh, no*. You thought you were just good friends, that it was a purely platonic thing. But, though you hate to admit even the possibility, are you really in love with your best friend? Does that mean that you have a duty to tell them how you truly feel, which could risk your friendship?

This sort of situation is not uncommon. Friendship and romantic attraction are not mutually exclusive. But to describe a relationship as platonic perhaps doesn't do justice to the origin of the idea. It can first be found in Plato's dialogue the *Symposium*, which takes place during a banquet attended by various eminent Greeks, including Plato's mentor Socrates.

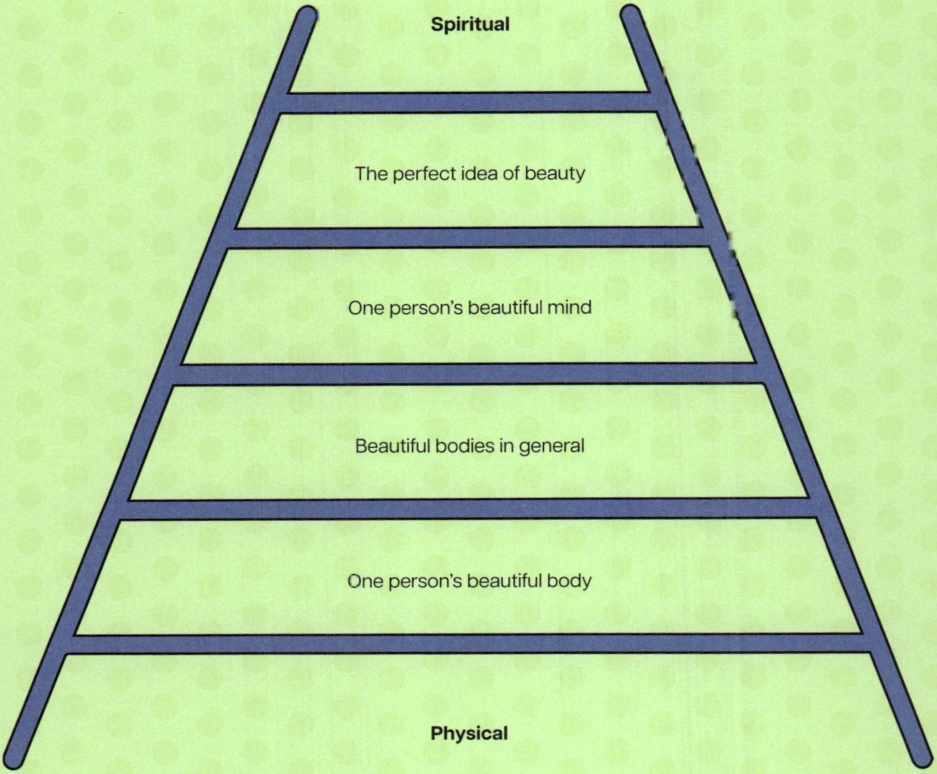


PLATO'S 'LADDER OF LOVE'

At some point, it is decided that all the guests will take it in turn to give a speech in praise of Eros, the god of love and sex. Each does so, emphasizing the different qualities of Eros. When it comes to Socrates' turn to speak, he recounts a conversation that he once had with the priestess Diotima, who argued that love is not a god, but a *daemon*, a spirit halfway between god and man. Therefore, the nature of love is more like a driving force that moves us toward the thing we consider desirable and beautiful. But, she says, there are different sorts of beauty. We may love someone's body, from which we may come to appreciate bodies, or the human form, in general. We may also come to love the mind or character – the soul – of the person we love and from that the beauty that resides in the thoughts, speech and actions of *other* people, even if their physical form is ugly (as Socrates himself famously was).

It is this progression – from the specific to the general, from the physical to the mental – that shows us how love can grow and evolve. Ultimately, then, love can bring us in touch with the divine ideal of absolute beauty (what Plato called its Form or perfect idea), and by doing so, elevate love to something spiritual and moral. For this reason, this vision of the potentially spiritualizing power of love is sometimes known as the ladder of Diotima.

To return to the situation with your friend, we can see that to call it 'just platonic' is not really accurate – unless, of course, your love for your friend *surpasses* physical or romantic love. In which case, it would surpass what most people think of as friendship, too.

That said, the ancient Greeks seem to have rated friendship very highly. In his *Nichomachean Ethics*, Aristotle devotes two whole books to the subject, considering friendliness one of the twelve chief virtues. He also argues that there are three types of friendship: the pleasant, the useful and the good.

A pleasant friendship is simply one where you enjoy each other's company. Perhaps this applies to you and your friend – you like the same books and films, after all. The useful friendship is more transactional. Perhaps your friend has a car when you need a lift, or maybe they're just a good listener when you want a moan. However, this needn't be one-sided: perhaps you fulfil similarly pleasurable or transactional roles in each other's lives.

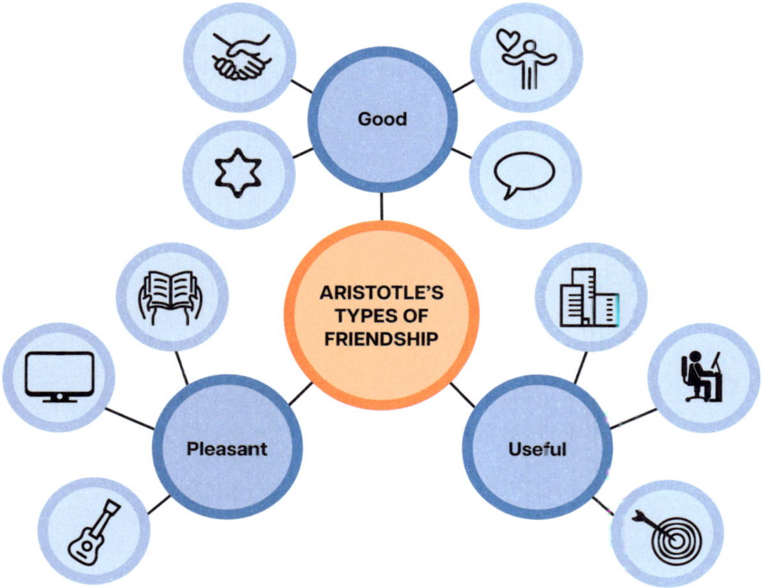


ARITSTOTLE'S THREE TYPES OF FRIENDSHIP

But true friendship is based not on a friend's pleasantness, nor their usefulness, but on their goodness. In other words, you enjoy mutual respect and admire each other's character. This means that a true friend might not always say what you want to hear, because sometimes – to quote English poet William Blake (1757–1827) – 'opposition is true friendship' (*The Marriage of Heaven and Hell*). Some hard truths might be just what you need to save you from a rash decision or a self-destructive act.

Other Greek philosophers agreed. Zeno, the founder of Stoicism, stated that 'All the good are friends of one another'. Epicurus thought it essential to human happiness and the epitome of the sort of moderate pleasure that he thought we should cultivate. Underlying all these views is perhaps a distrust of sexual attraction and the passions of the flesh. It was common among the ancient philosophers to caution against the disruptive role that bodily desire could play, unseating reason and steering us toward impulsive behaviour.

Where does all of this leave you in your dilemma? I'm not an agony aunt or a psychologist, so it's maybe not my place to advise. But – speaking as a 'friend' – I think you need to be honest with yourself. Aristotle would ask, are you behaving like a true friend? If to tell your friend how you really feel risks not only your friendship but their new relationship too, then are you behaving selfishly? In the play *Cyrano de Bergerac* (1897) by French dramatist Edmond Rostond (1868–1918), Cyrano writes love letters to Roxanne on behalf of her tongue-tied suitor, Christian. The letters profess a deeper, more profound love than the fairly basic but good-looking Christian could imagine. When Roxanne rumbles the ruse, Cyrano denies it, almost to the point of death. So, is your love more fully expressed by keeping silent about it?

The deepest love is altruistic and keeping your possibly romantic feelings to yourself – especially if your friend doesn't reciprocate them – may be a higher expression of love than the romantic variety. If that's how you truly love your friend, then nothing can get in the way of it.

# THE WORLD AND MY PLACE IN IT

## POLITICAL PHILOSOPHY

5.

# CAN I HAVE SOME PRIVACY, PLEASE?

**I just got a notification from my fridge: apparently, I'm low on milk. But not to worry, because the appliance has already added it to my online shopping list. I'm in town, and my smartwatch pings to tell me that a friend is nearby: do I want to message them to see if they'd like to catch up over coffee? A website emails me to tell me that there is now a sequel to the book I recently purchased: do I want to buy it? I can unlock my phone with my face and my laptop with my fingerprint, and an app tells me exactly where my daughter currently is (not in fact studying at a friend's house, like she'd told me, but hanging out in the local shopping centre).**

An increasing proportion of our lives is now online and it's possible to build up a detailed picture of a person from the data amassed by their shopping, romance, work and leisure activities. But at what point does the collection of all this information move from being convenient to creepy?

I guess it all depends on who has access to it: governments, parents, corporations, law enforcement agencies, criminals, your employer. Whatever the case, it seems that many countries are edging toward the sort of total-surveillance society envisaged by George Orwell in his novel *1984*. At least in terms of the scope and detail of the information that is potentially available to the authorities. Collection of this information seems legitimate if it's used to fight crime and terrorism or to provide civic services and healthcare. But at what point does this erosion of privacy become illegitimate? When does Big Data become Big Brother?

The French philosopher Michel Foucault (1926–84) was especially interested in the subtle ways in which power limits and shapes our lives. Writing pre-internet, he was nonetheless ahead of his time in outlining how the invasiveness that technology now allows could be wielded by those who control the form and flow of information. In the fields of education, mental health and criminal justice, Foucault tried to show how our lives are as much shaped by words and concepts as by any physical forces or deterrents.

By defining who is mad and who is sane, what is legal and what is criminal, power shapes society's attitudes and values in relation to what is and isn't acceptable. Moreover – and here is the subtler point – in being shaped by these processes, society's members also internalize them, so that it becomes less necessary to monitor and surveil citizens in order to enforce order. Instead, society polices itself.

In *Discipline and Punish* (1975), Foucault further illustrates this point through discussion of the panopticon, a prison designed by the philosopher Jeremy Bentham, but never built. Imagine a circular building constructed of many levels, on each of which are rows of cells. The cells are open-fronted, with only three walls, so that the activities of their inhabitants are visible at any time from a guard tower situated in the centre of the prison. However – and here is the stroke of genius – the observation tower is constructed in such a way that the inmates cannot see whether there is actually anyone staffing it.

In other words, they never know if or when they are being watched. As a result, they assume that they are and effectively do the job of the prison staff for them: they watch themselves. And so, in a similar way to the inmates, Foucault argues that, by the internalization of the values, norms and concepts used by power to classify behaviour, the members of a society are educated and conditioned to monitor, judge and restrict their own actions, thoughts, feelings and attitudes.

In *The History of Sexuality, Volume 1* (1976), Foucault discusses perhaps the most powerful illustration of this process, the Catholic confessional. Here, believers were encouraged to catalogue their sins and mental misdemeanours to the priest, who would remind them that these were already known to God, who can see into their deepest hearts. Thus, the believer carries on their day-to-day life with a perpetual sense of being watched and behaves accordingly.

In the case of modern technological communication and the internet, we can witness a literal enactment of this relationship, replacing God or the guard tower with the authorities, who at any time could search your emails, texts, browsing history, even perhaps covertly spy on you through your phone's microphone or your laptop's camera. In the meantime, you are encouraged to reveal your innermost thoughts on social media, to share your private and personal views, your sexual preferences, your sporting and political affiliations, your taste in books, music, food and fashion.

THE OPEN PRISON

It's reached a point where those who prefer *not* to share are almost viewed with suspicion. You're not on Facebook? What's wrong with you?

Possibly, this is all paranoia. Just because the government *could* access your private information (should it have legitimate cause), doesn't mean that it *is*. If you have nothing to hide, then what's the problem?

But of course, we all have things to hide – even if it's just that embarrassing photo of you when you were twelve and your mother had persuaded you to get That Haircut. Surely, we have a right to some privacy, don't we? Well, there have been movements to address this, such as the privacy that is enshrined in the UN's *Universal Declaration of Human Rights*, and the EU's general data protection regulation (GDPR) legislation, which came into effect in 2018.

We have to wonder if such things provide sufficient protection against the overreach of governments and the insidious monitoring by the tech companies themselves. For, having sleepwalked into a world where your fridge now knows more about your food preferences than your mother does, have we created a technological state apparatus that is ripe for misuse by political bad actors?

# FAIR PLAY OR FAIRY TALE?

**It is some years in the future. Earth is dying and humanity decides to make a new home on Mars. However, with so many cultural differences, faiths and practices, what values and principles should form the basis of the new society in order to ensure justice and equality for all Earth's people? And how are we to agree upon them?**

In *A Theory of Justice* (1971), American philosopher John Rawls (1921–2002) argued that an ideally just society should be primarily judged on how fair it is to all members in terms of its basic structure and institutions. To arrive at such a society, he proposed a thought experiment: imagine that in the new colony, you do not know what status or personal qualities you are to possess. In other words, you don't know if you're to be rich or poor, tall or short, black or white, male or female, atheist or believer, able-bodied or disabled. You would be, in effect, making decisions from behind a veil of ignorance, as Rawls terms it, and would be motivated to establish laws and institutions that benefit you no matter who you turn out to be.

One rational strategy such a quality-less person might adopt, Rawls argues, is a maximin approach, whereby they try to ensure that society maximizes the welfare of those who are least well-off so that the worst possible outcome is minimized. For example, rather than allowing the rich to amass limitless wealth while the poor starve, a just and fair society would permit a degree of inequality, but redistribute excess wealth through taxation, thus protecting the poorest and ensuring their basic needs are covered.

This hypothetical situation, or 'original position', as Rawls terms it, might seem like an artificial and perhaps impractical approach, almost a philosopher's fairy tale. And yet, such a thought experiment has been a common tool for political philosophers over the years, who have similarly imagined the existence of humanity in a raw state of nature, prior to society or government, from which they could propose how a viable social contract might emerge between society's members and the state. It's interesting to note how different these answers have been.

For instance, English philosopher Thomas Hobbes (1588–1679) argued that humanity is primarily driven by self-interest and, in a state of nature, people would not hesitate to murder, bully and exploit others for their own ends. Hobbes suggested that the main role of government was therefore to protect its citizens from each other, and to do so, it must wield absolute authority and power to enforce laws by whatever means necessary.

In contrast, French philosopher Jean-Jacques Rousseau (1712–78) had a much more optimistic view of human nature, believing in our innate compassion and goodness. And while he believed that civilization had an ultimately corrupting influence, he envisaged that this could be offset by democratic means as society's members came together to ensure equality and mutual freedom through the expression of the collective will of the people.

But how feasible is Rawls's thought experiment? I might be able to imagine, to an extent, what it's like to have no money or lots, or to be a different race or sex, but can I really

just shelve all my inherent biases and deep-seated values? Furthermore, would I want to? An alternative approach is known as communitarianism. This argues that we shouldn't imagine ourselves divested of all pre-existing qualities when trying to establish a common ground for our new society but should look to those values that each of us already possesses by virtue of our distinct cultural backgrounds.

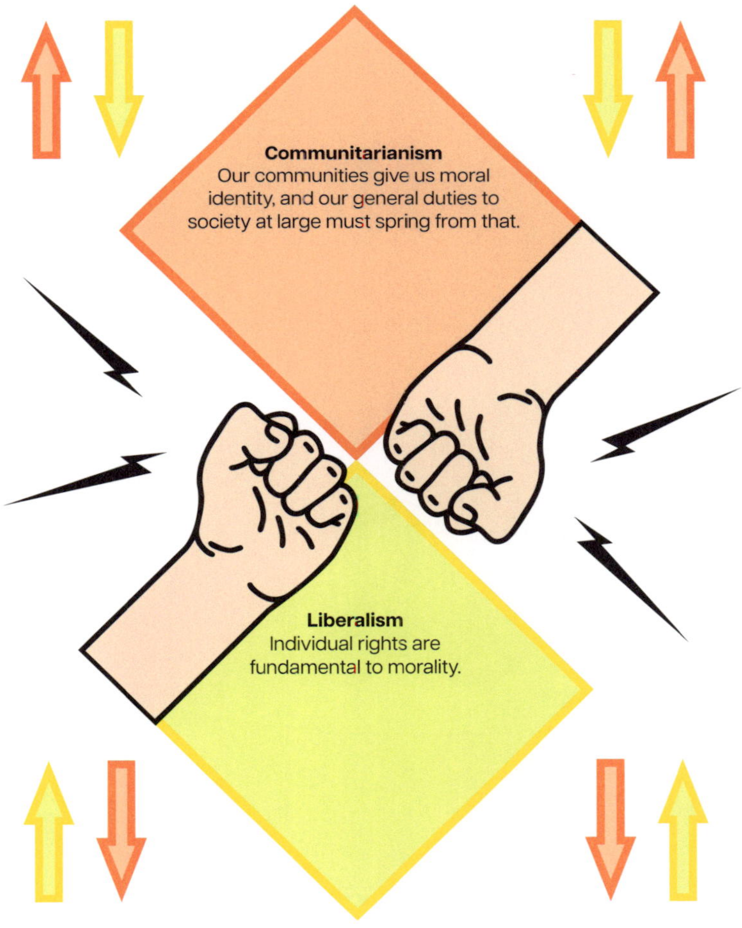


THE COMMUNITY VS THE INDIVIDUAL

This would mean that rather than imposing an artificial and abstract approach from the outside, we would be better off identifying the common values held *inside* Earth's various cultures. Most communities want to ensure the safety of their members, to educate their young and care for their sick and elderly. So, no matter what other cultural differences exist, the core basis for wider social cohesion already exists in the personal history, upbringing, inherited values and beliefs of each of its citizens. Communities give us that, not adherence to some overriding external rational principle.

That said, the communitarian approach can be problematic. Let's say that a certain group within a wider society forbids homosexuality or the free choice of a marriage partner. Should the society as a whole simply accept that, or does the state have a duty to step in and support *what it sees* as the rights of the individual within that minority group? If it doesn't, out of respect for that community's own internal values, this risks fracturing the overall society of which the community is a part. These fractured parts may then form into isolated groups with their own values. All of which seems like a step towards moral relativism, or the idea that there is no such thing as universal moral values of the sort Kant argued for (see page 92).

To return to our new Martian colony, we can easily see how the respect of such cultural differences may lead to conflict or even war. This is one of the reasons that Rawls proposed the veil of ignorance in the first place, because we wouldn't know in advance which community we would be a member of, whether we would be gay or straight, rich or poor, or whatever. In this sense, as artificial and arbitrary as it might seem, Rawls's approach may in fact be the more practically workable, and not so much of a fairy tale after all.

# IS THE WORLD GETTING BETTER OR ARE THINGS JUST DIFFERENT?

**We are used to thinking of history as progressive, but it hasn't always been this way. For instance, the Abrahamic religions of Judaism, Christianity and Islam all see humans as having declined from an initial state of blissful existence in the Earthly paradise of the Garden of Eden. And since it was a selfish action – a sin – that caused this fall, then reclaiming that state is usually seen not in terms of the application of intelligence, skill and effort, but rather of renewed faith and trust in God.**

Similarly, in his *Works and Days*, the Ancient Greek poet Hesiod (born *c.*700 BC) describes the descent of humanity from its earliest Golden Age of peace, happiness and plenty, through five declining ages, each less ideal than the last, culminating in the Iron Age of his own time, where the gods had abandoned humans to their own selfishness and corruption. A similar scheme was later employed by the

THE PROGRESSIVE VIEW OF HISTORY

Roman poet Ovid (43 BC–AD 17) in his *Metamorphoses*. In broad terms, such views are called prelapsarian, in that they refer to a time before humanity's fall or lapse from a perfect existence.

The opposite view, which sees humans as developing out of a primitive state, may be termed evolutionism. Just as English naturalist Charles Darwin (1809–82) argued that humans evolved from more primitive biological forms, so we might also use the term more loosely to imply our progress in social and cultural matters. This view of history is derived in part from the Enlightenment. During this period in the 17th and 18th centuries, impressive strides were taken in science and a greater emphasis was placed upon reason and the rights of the individual. As thinkers such as Isaac Newton, Francis Bacon (1561–1626) and René Descartes replaced superstition and faith with rational explanation and scientific method, so philosophers including John Locke, Jean-Jaques Rousseau and Adam Smith (1723–90) attempted to provide a rational basis for political government and economic activity.

When such liberal ideals inspired the French Revolution in 1789, it was more than just the monarchy that was overthrown. Traditions, customs, laws and attitudes that embodied the old class-based way of thinking, which saw people as more or less confined to the place God had allotted them by birth, were also toppled. However, through our own endeavours, the revolutionaries argued, it was now possible

THE PROGRESSIVE VIEW OF HISTORY

to see humanity as collectively evolving toward a world where all would be equal, regardless of birth or wealth, sex or religion.

Similar ideals influenced the 1917 Russian Revolution. The German philosopher Georg William Friedrich Hegel (1770–1831) argued that history was a dialectical process, whereby human society slowly evolved by lurching between opposing ideals, almost as if arguing with itself. This meant a period of strict authoritarianism might give way to a period of extreme liberalism, which could in turn influence a return toward social order that represented a compromise between both extremes. The process would then begin again with a new starting point.

This perpetual movement between one extreme (thesis), its opposite (antithesis), and a resulting compromise (synthesis), would eventually lead to the final establishment

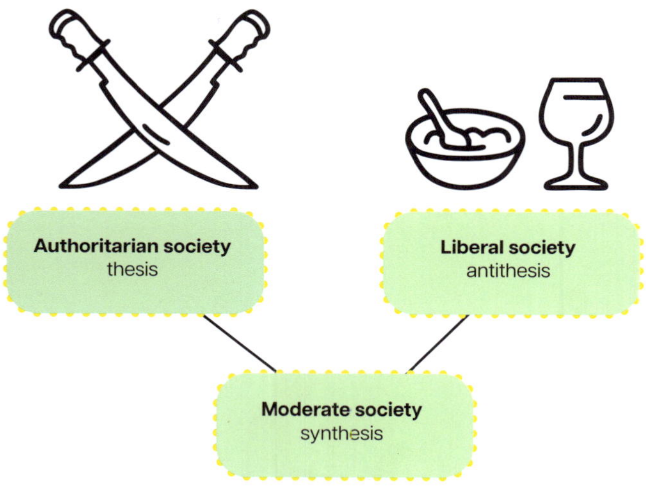


THE HEGELIAN DIALECTICAL MODEL

of the ideal political state, where people would live free, ethical and rationally guided lives. This view influenced the communist philosopher Karl Marx, who predicted that the same material forces that had driven primitive communal societies to evolve through social classes, slavery, feudalism and eventually into modern capitalism, would also bring about the ideal communist utopia.

At least, that was the theory. In practice, both the French and Russian Revolutions spawned infighting, corruption and an idealized concept of human nature. The 20th century dealt a heavy blow to the progressive view through the horrors of two World Wars, the atomic bomb and the Holocaust, but the onward march of science and technology has kept the premise alive. And it cannot be denied, viewed from the long perspective, that things are looking up. People are, in general, living longer, healthier lives. Infant mortality rates have greatly decreased and there are fewer people in poverty or dying from disease. Compared to other periods of history, there are – believe it or not – fewer wars and fewer people die from violence. In social and political terms, people are also generally freer and more equal and have better access to education and healthcare, with higher standards of living, in most countries, at least.

It is such achievements that led American political scientist Francis Fukuyama (1952–) to declare, in his 1992 book, *The End of History and the Last Man*, that history was over. With the fall of the Berlin Wall in 1989, capitalism and the Western way of life had won, and if there was an endpoint to history – as both Hegel and Marx thought there would be – then it was one based on free-market economics and liberal democracy, not on state-managed communism.

Which all sounds great. Except of course that, as current events continue to reveal, history doesn't seem to have received the message. Despite the onward march of science, the promised technological utopia has not emerged, modern life is rife with anxiety and overwork, and social media is creating a mental-health epidemic among our young people. Nationalism is on the rise, significant military conflicts play out around the globe and with the advent of AI and automated warfare, science and technology seem as much of an existential threat as the problems they are meant to solve. Inequality is also increasing, with – according to a 2024 report by the charity Oxfam titled 'Multilateralism in the Era of Global Oligarchy' – 1% of the global population owning more wealth than the bottom 95%, while the streets of the wealthiest cities remain blighted with homelessness and poverty. Oh, and the planet might be dying.

Does this seem like progress? Well, yes and no. What progress gives with one hand, it would seem to take away with the other, and while we have indeed made great strides, these have also been accompanied by the advent of fresh problems. Personally, I can't help wondering, all this progress aside, whether we are truly happier than our ancient ancestors. To strike an optimistic note, perhaps this is the wake-up call that we need, and, if Hegel is right, history has one more correction to make.

# IS FIGHTING ALWAYS WRONG?

**The inhabitants of most countries – especially those who have witnessed first-hand the true horrors of war – would prefer to live peacefully with their neighbours. However, a number of philosophers have recognized that, as repugnant as war is, there are rare and specific circumstances in which it may be morally justified.**

The beginnings of the theoretical discussion of the ethics of war can be traced back at least to the Roman Empire, where the great orator Marcus Cicero (106–43 BC) was among the first thinkers to outline the concept of *bellum iustum* (just war). In *On Duties* and other works, Cicero argues that war should be considered a last resort. If military action is being considered as a response to another country's aggression or wrongdoing, then has an official attempt at least been made to resolve the situation peacefully? Has a demand for recompense or apology been made? Have any negotiations been attempted? It's only after such civil avenues of action have amounted to nothing that more strident means should be considered.

And even then, it should be clear that military action is solely necessary to regain honour and deter further aggression against one's country. War should be declared only by a proper authority – it would not be justifiable for a man to raise his own private army in order to settle some personal score or plunder riches. Fundamentally, the aim of the war – once justice and deterrence have been served – should be peace.

Taken together, such considerations are traditionally termed in Latin the *ius ad bellum*, or justification for war. Even so, Cicero argues, once these conditions have been met, we

must behave according to the dictates of *ius in bello*, or just conduct within war – that is, the observance of ethical and humane behaviour.

These principles were later more fully developed by Christian theologians such as St Anselm (*c.*1033–1109) and St Thomas Aquinas (1225–74). They accepted many of Cicero's points, while simultaneously trying to square the occasional necessity of war with the teachings of Jesus, an absolute pacifist who had preached 'Love your enemies' (Matthew 5:44) and 'Whoever strikes you on your right cheek, turn to him the other also' (Matthew 5:39). Aquinas justified war via his doctrine of double effect. This is when actions have both good and bad consequences: I protect my own life or that of others (a good action), but in the process I take my attacker's life (a bad action). Such actions are justified only if four criteria are met:

**1** Our intentions are good or at least neutral (for example, we were acting in self-defence or just going about our daily business).

**2** We intend only to do good and the bad is an unfortunate possible side-effect of intending the good action (you repel the attacker in order to survive; his death is a foreseen but regrettable possibility).

**3** The outcome of the good action must be direct and the bad action not an indirect means to achieve the good. So, a long-term strategy of targeting innocents (bad) might eventually end a war (good), but that doesn't justify the bad act.

**4** There is a proportional and serious reason for the bad effect (there must be a strong reason to go to war, but if it could be done another way, with fewer or no bad consequences, then that way should be preferred).

When we look to modern times, we can see that many wars do not meet these criteria. In a modern update of Aquinas's double effect, powerful long-range weapons and bombs mean that the death of civilians is now often considered collateral damage when targeting an enemy's infrastructure, or even as a legitimate means of inducing terror and destroying morale. Countries have historically made use of engineered pretexts, such as the use of false flag operations (where an act of aggression is falsely blamed on a foreign country's agents), or the assertion that military intervention is required to rescue a foreign country from its own dangerous elements.

Underlying such callousness and immorality is a different opinion as to the point of war, which is in turn related to a darker view of human nature. As noted earlier (see page 162), English philosopher Thomas Hobbes had a pessimistic view of humanity and he considered unregulated human relations to be 'a war of all against all'. Hence, he argued, an all-powerful government was required to ensure peace.

But when applied to global politics, where no such authority might exist, this view explains how each nation is driven by its own inherent need to pursue dominance and self-interest. In the context of international relations, Hobbes's position has become known as political realism and was memorably articulated by the Greek historian Thucydides (*c.*455–*c.*400 BC) in his *History of the Peloponnesian War*: 'The strong do what they can and the weak suffer what they must.' In other words, when you can get away with it, violence is often considered the most effective

THE EVOLUTION OF WEAPONS

means to achieve what you want. A similarly bleak view of human nature was espoused by the Italian political philosopher Niccolò Machiavelli (1469–1527) in *The Prince*, posthumously published in 1532. This quote sums up his position nicely:

> 'There are two methods of fighting, the one by law, the other by force: the first method is that of me, the second of beasts; but as the first method is often insufficient, one must have recourse to the second.'
>
> — Niccolò Machiavelli, The Prince

But must we behave as beasts? And must we predict that our enemies will also do so? Machiavelli's advice to rulers was that they should assume the worst about people. And yet, while the history of war supplies more than enough evidence to back this up, it also supplies a great deal in support of the contrary view, of examples of selfless heroism, of those who risk their lives to save others or sacrifice it in the name of peace. Even if human brutality far outweighs acts of saintliness in terms of sheer numbers, shouldn't the latter serve as evidence of possibility and of what we might strive to become?

# CAN I SAY ANYTHING I WANT?

**You are a keen amateur gardener. It comes to the attention of the local press that you have managed to grow a rose of extraordinary size. Soon, the story goes viral on social media, and there is international interest from newspapers, radio and TV. During all the hubbub, in an unguarded moment, you happen to say something controversial. Suddenly, the mood changes and focus shifts away from your enormous flower and onto you. People turn against you. You start receiving abusive messages, even death threats. That escalated quickly! But what did you do to deserve it? Don't you have a right to freedom of speech?**

We live in an age of so-called cancel culture, where a verbal misstep can land you in the court of public opinion, a throwaway comment can occasion a social-media pile-on, and an off-colour joke can result in the loss of your job. Yes, of course, you can say what you want within the bounds of the law. Most countries enshrine some form of freedom of speech, and the United Nations' *Universal Declaration of Human Rights* considers it a fundamental freedom. But this cuts both ways and it also means that others are free to voice their opinions on what you've said and the sort of person they think you are. But there must be some limits to this, surely? I mean, *death threats*?

In *On Liberty* (1859), a foundational text of liberalism, John Stuart Mill argued that the only limit to freedom of speech should be if it causes harm to another. What should this harm consist of? For Mill, it covered bodily danger, libel and slander. If someone posts on social media something like:

then it might be considered incitement to violence and therefore a criminal offence. If someone writes (libel) or speaks (slander) about you in such a way that it causes significant harm to your reputation, then this might also be problematic for them – unless it's true. If I say, 'This man is a thief and a liar', and it does indeed turn out that you have stolen and lied, then that's not defamatory, even though you might not like to hear it.

Different countries have different regulations concerning freedom of speech. For instance, the UK is considered more claimant friendly, because the burden lies with the defendant to show that their words are not libellous or slanderous. In the US, where freedom of speech is enshrined in the Constitution, the burden is on the claimant to show that they have been harmed by another's words. This can lead to what is termed libel tourism or forum shopping, where foreign nationals will try to bring defamation cases in UK courts.

In some countries, truth is not a complete defence and defamation can still apply if the remarks are considered harmful to a person's reputation, or if what is said is not deemed to be in the public interest. But let's say that you live in a country where truth is a complete defence and your statement is not libellous, slanderous nor an incitement to violence, merely an unpopular opinion among a certain vocal percentage of the population – even, let's say for argument's sake, most of it. Do the offended people have a right to silence you? As already noted, this is termed by Mill 'the tyranny of the majority', whereby the views of minorities are silenced (see page 70). This can happen informally, through the expression of public disapproval through the media, or legally, through the passing of discriminatory laws favoured by the majority.

Most governments now afford better protections for minority views, but – as evidenced by cancel culture – this does not protect you from the ire of the masses (assuming they don't resort to libel, slander or death threats). So, should there be laws that stop you from offending others and from them offending you?

The problem here is that offence is difficult to define. What offends one person might not offend another, so the standard is subjective. An exception is hate speech, which more broadly protects individuals against such

Mill thought the only limit to freedom of speech should be if it causes (or incites) actual harm to others.

THE PRINCIPLE OF HARM

discrimination and prejudice based on certain protected characteristics, such as race, ethnicity, nationality, religion or philosophical belief, gender, sexual orientation and disability. So, what you say might not be libellous, slanderous or incitement, but if I insult you or behave prejudicially toward you based on one of these protected characteristics, then that might be against criminal or civil law. However, once again, these laws differ by country and we see a difference between the US and other Western democracies. In America, even hate speech may be tolerated as a Constitutional right, unless it poses a direct threat.

For Mill, however, legislating for hate speech that does not incite harm would be going too far. If we suppress or punish extreme views, then we lose the opportunity of engaging with them, and thereby demonstrating by rational argument how they are wrong. Or, the extreme view might be correct, and the majority thereby loses out on knowing the truth if it's suppressed. Either way, society is healthier if we tolerate a range of views, as long as we engage with them.

There are various problems with this. Among them is the argument that a failure to suppress bigoted and prejudicial views can allow them to gain traction. This is especially true in the age of social media, which is undermining our ability to engage in a rational and measured way with the things we are exposed to. Arguably, this can directly harm young minds who are not yet fully equipped to process such views in a mature and considered way.

It also risks what Karl Popper termed the paradox of intolerance, whereby the toleration of views that themselves are intolerant undermines the practice of tolerance. In other words, if intolerance is allowed to spread, then a government might be voted into power that then rolls back laws promoting tolerance. Whatever the case, it's a thorny issue.

You might be best sticking to your gardening.

# IS IT OK TO BE LAZY?

**It's the weekend. You should be putting your feet up, spending time with the family, maybe doing a bit of light pottering in the garden. But instead you're on your laptop or your phone, answering work emails, networking on social media or working on that presentation for Monday. 'Just tying up a few loose ends,' you explain to your partner as you disappear off into your home office. But your partner's busy grading homework assignments at the kitchen table and doesn't hear you.**

What happened to hobbies, or to time off? Didn't there used to be such a thing as work–life balance? When did 'leisure' become a dirty word? If we are to believe German sociologist Max Weber (1864–1920), it was sometime in the 16th century, during the Protestant Reformation and the advent of Puritanism. French theologian John Calvin (1509–64) preached that God's grace could not be earned, but he also argued that Christian believers could manifest their faith through 'good works', so the pursuit of a profession could be seen as the fulfilment of a divine 'calling'. Consequently, 'busyness' itself became a virtue and worldly success an indication of God's favour, while inactivity was looked upon with mistrust.

This dovetailed nicely with the growth of capitalism and the needs of industrialization, as rural workers were encouraged to transition from a lifestyle that often involved long periods of downtime to one that required fixed daily work schedules. Gradually, then, leisure became associated with laziness, with guilt, with not pulling your weight, and – through that catchy proverb, 'The Devil finds work for idle hands to do' – with sin.

The irony is that even in countries where religious belief has declined, the Protestant work ethic persists. It's almost as if it has replaced some religious need. 'Hustle culture' is now a global phenomenon, the latest manifestation of which is the 996 working hour system – working from 9am to 9pm, six days a week – which originated in China but is increasingly being adopted by Western tech and finance companies. Even the introduction into the business world of the practice of mindfulness, a secular borrowing from Buddhism that is meant to increase mental health, is often seen as a way of increasing productivity and staving off burnout, as if the real purpose of opening your third-eye chakra is to enable you to hit your third-quarter sales targets.

When Karl Marx dreamed of a communist utopia, he envisaged that by working together to meet the community's needs (and not fill capitalist pockets), we would naturally have more leisure time. However, given that Marx also thought productivity a fundamental aspect of human nature, he saw the main difference between capitalist and

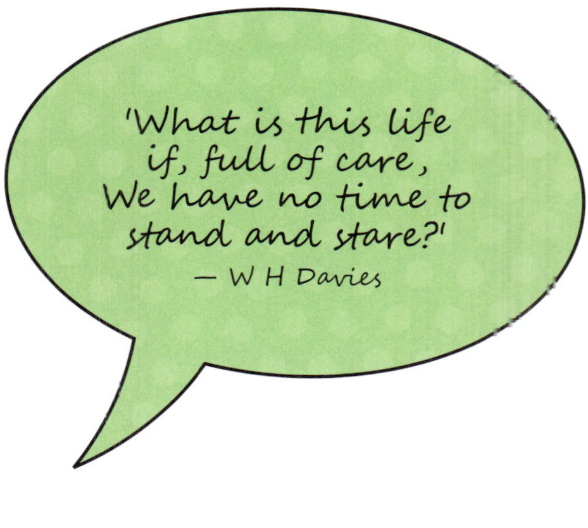

communist culture as consisting in choice. Famously, he pictured a day where he might go hunting in the morning and fishing in the afternoon, then do a bit of cattle husbandry in the evening and write some literary criticism after dinner. For Marx, therefore, leisure lies not so much in inactivity, but in choosing how we spend our time.

Taking a more individualist and independent outlook, in his 1854 book *Walden*, American transcendentalist writer Henry David Thoreau (1817–62) details his two-year experiment of living alone in a cabin in the woods by Walden Pond, in Concord, Massachusetts. Here, stripped of many of the luxuries of civilized life, Thoreau does indeed spend his time hunting and fishing, reading and writing. He is not completely isolated – he makes weekly trips into Concord for provisions and to catch up on the news, and he occasionally spends time with friends – but this semi-detachment from the bustle of civilization allows him to connect with nature and to reflect on the purpose of human life.

It might be nice to wander through the woods watching squirrels and foxes, but perhaps the thought of having to catch or grow your own food feels a bit *too* much like work. Can't we strike a balance? What about technology? Isn't the whole purpose of automation to free us up to do what we please? Marx thought so, as did Aristotle, who with remarkable prescience imagined a time where automated tools would do their own work, liberating us from manual labour to pursue more noble activities (such as philosophy!).

It is true that today's average employee works fewer hours than their nineteenth century counterpart, and has more holidays and time off. However, as we enter an age with technologies that have the promise to deliver on the dreams of a life of leisure, the modern workforce is also coming under pressure to deliver more.

With the increase in remote and flexible working, it's become harder to divorce 'home' from 'work'. Add to that the downside of internet connectivity and 'always-on' culture, and it's easy to see how despite our apparently better working conditions, workers are more and more likely to report work-related stress, an inability to switch off and general discontent. And the advent of AI, while also endangering careers and livelihoods, has only seemed to ramp up expectations regarding productivity: we are expected to do more, quicker and with less.

So what do we do? Marx would have said that the problem is *alienation*, or the fact that most of us are forced to work for a wage doing things that aren't directly related to our needs. We are insurance brokers, florists or writers of popular philosophy not because those activities directly meet our survival needs or leisure interests, but because they serve to generate profits for the business owners and keep the economy going.

But if AI and automation are *not* the answer to this, then what is? The trend of 'quiet quitting', or doing the minimum that your job requires, seems like a meaningless personal protest. What we need is a completely different attitude to work and what it's for.

A sort of revolution, perhaps.

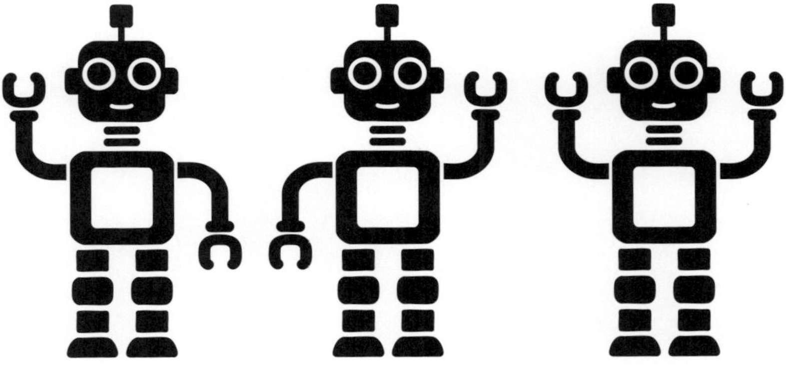

# INDEX

# ABOUT THE AUTHOR

**GARETH SOUTHWELL** is a writer, illustrator and philosopher who lives in South Wales in the UK.

He's been involved with philosophy for over thirty years, is the author of numerous books on the subject, and has a PhD on the topic of 'death' (which has been read by exactly four people, including himself).

His sci-fi/fantasy novels are just a thinly fictionalized version of his PhD thesis, with most of the boring bits taken out and replaced with robots and people swearing in Welsh.

Aside from all that, he is a professional illustrator and book designer, and runs WoodPig Press, a micro-press specializing in speculative fiction and non-fiction that celebrates human creativity and opposes all things AI.